创意无限：
Midjourney AI商用绘画300例
（120集视频课）

>>>罗巨浪 著

清华大学出版社
北京

内容简介

人工智能时代，AI绘画作为AIGC技术的一项重要应用，正在改变着艺术创作的方式，Midjourney、Stable Diffusion、文心一格等AI绘画工具的出现为艺术家带来了新的创作工具、创作思路和创作灵感，为艺术领域带来了全新的机遇和挑战。本书就是一本通过Midjourney创作的AI绘画案例书，涵盖插画绘制、DIY设计、手办设计、LOGO设计、海报主图设计、自媒体相关设计、摄影效果制作、装帧及文创设计、展品设计和室内设计等常用商业设计领域的300多个绘画案例。每个案例都是按"绘画主体＋场景＋风格＋画质＋基础设置"的模块组织关键词，改变各模块的关键词，即可得到新的绘画作品。

本书采用全彩印刷，案例丰富多样，适合所有AI绘画爱好者学习，专业设计师也能从本书中获得创意灵感，拓展创作思路。另外，本书也非常适合作为相关培训机构的教材，帮助学员在AI绘画领域快速进步。

版权所有，侵权必究。举报：010-62782989，beiqinquan@tup.tsinghua.edu.cn。

图书在版编目（CIP）数据

创意无限：Midjourney AI商用绘画300例：120集视频课 / 罗巨浪著. -- 北京：清华大学出版社，2024.7. -- ISBN 978-7-302-66708-7

Ⅰ．TP391.413

中国国家版本馆CIP数据核字第20246NT211号

责任编辑：袁金敏
封面设计：墨　白
责任校对：徐俊伟
责任印制：宋　林

出版发行：清华大学出版社
网　　址：https://www.tup.com.cn，https://www.wqxuetang.com
地　　址：北京清华大学学研大厦A座　　邮　编：100084
社 总 机：010-83470000　　邮　购：010-62786544
投稿与读者服务：010-62776969，c-service@tup.tsinghua.edu.cn
质 量 反 馈：010-62772015，zhiliang@tup.tsinghua.edu.cn
印 装 者：天津鑫丰华印务有限公司
经　　销：全国新华书店
开　　本：170mm×240mm　　印　张：21.5　　插　页：1　　字　数：551千字
版　　次：2024年7月第1版　　　　　印　　次：2024年7月第1次印刷
定　　价：128.00元

产品编号：107048-01

前言

AI（Artificial Intelligence，人工智能）的历史意义可与互联网的诞生相媲美，如比尔·盖茨所言，它标志着一个全新时代的到来。一本英国知名新闻和商业杂志在一篇文章中指出，AI引发了一场可比肩蒸汽机、电力和计算机时代的巨变，它是第四次工业革命的核心。在AI绘画领域，这种变革尤为显著。AI绘画软件通过简单的命令输入，能够在极短时间内创作出高质量的作品，满足了插画艺术、广告设计、出版传媒和电商等多个行业的需求，从而将这一技术转化为实际工作中的常规工具。

然而，要有效利用AI绘画软件并充分发挥其潜力，专业的训练和实践是必不可少的。掌握了AI绘画技术，即使是初学者也能迅速提升技能，成为熟练的插画师或专业设计师。适当的训练和实践不仅能够帮助初学者相信AI绘画的有效性，还能够充分利用这项技术，为工作和创作带来极大的便利。

为了满足广大读者的学习和实际工作需求，我们创作了《创意无限：Midjourney AI商用绘画300例（120集视频课）》。本书具有以下四大特色：

场景式学习：根据工作和生活中的实际需求，特别是商业需求，我们设计了多种使用场景，如插画绘制、DIY设计、手办设计、LOGO设计、海报设计、自媒体相关设计、摄影效果制作、装帧及文创设计、展品设计和室内设计等。读者可以根据自己的需求，对不同场景下的案例进行调整，实现"所学即所用"。

案例式学习：本书展示了300多个案例，重在实操。读者可以通过替换案例中的提示词，创作出不同风格的新作品。这种方法可以帮助读者快速掌握AI绘画技巧，做到举一反三。

"傻瓜式"学习：所有AI绘画案例的创作过程都被分解成多个简单明了的步骤。无论读者是否具备绘画基础，都能按照教程创作出属于自己的AI图像。

"万能绘画公式"的使用：为了帮助初学者更好地描述想要的画面，本书介绍了"万能绘画公式"：绘画主体＋场景＋风格＋画质＋基础设置。读者可以根据自身需求，任意调整"万能绘画公式"中的提示词，生成自己的绘画"公式"。这种方法既灵活又实用，特别适合在创作过程中希望调整元素和风格的读者。

无论你是初学者还是有经验的读者，我们都希望本书能成为你学习和应用AI绘画技术的得力工具。

为了更好地帮助读者，本书特别录制了120集关于Midjourney绘画软件的使用技巧、重要参数及命令的介绍视频,并特别赠送了300多个案例的提示词及效果图(电子书),以及关键词速查表、艺术风格与流派的思维导图,读者可根据需要扫描下面的二维码,下载使用!

致　谢

衷心感谢李虹霖、黄燕平、周冰渝、白朋、王鹏飞、王晓铃在图书写作、版式设计、图书装帧中给予的帮助。

衷心感谢所有编辑、校对等人员对图书内容的精心审读、加工,使其符合出版规范。

针对本书的出版,尽管已付出了很多努力,但限于时间、篇幅,也难免存在疏漏之处,敬请广大读者批评、指正,不吝赐教。

——编者
2024年7月

目录

第 1 章　Midjourney 入门 ——— 001

1.1　Midjourney 的安装与订阅　002
1.2　用文字生成图像　009
1.3　出图界面常用按钮介绍　010
1.4　打开和保存图像　014
1.5　重绘按钮和 Zoom 按钮　014
1.6　设置面板的基本功能　020
1.7　常见参数　024
1.8　图生图的使用方法　025
1.9　多张图片混合生成图像的方法　027
1.10　反推提示词（/describe 命令的使用）　029
1.11　万能绘画公式的使用　031
1.12　文字的使用　034

第 2 章　插画绘制 ——— 035

例 1.　插画基础　036
例 2.　线稿插画　037
例 3.　油画风插画　038
例 4.　国画风插画　039
例 5.　敦煌风插画　040
例 6.　人物设计插画　041
例 7.　木刻版画　042
例 8.　彩色玻璃插画　043
例 9.　卡通插画　044
例 10.　像素人物插画　045
例 11.　游戏场景设计插画　046
例 12.　赛博朋克插画　047
例 13.　涂鸦插画　048
例 14.　彩色铅笔插画　049
例 15.　漫画分镜插画　050
例 16.　水彩插画　051
例 17.　乡村风景插画　052
例 18.　扁平风格插画　053
例 19.　穆夏风格插画　054
例 20.　宫崎骏风格插画　055
例 21.　哥特式风格插画　056
例 22.　街景图片插画　057
例 23.　日式漫画风格插画　058
例 24.　卡通风格插画　059
例 25.　抽象线条风格插画　059
例 26.　简笔插画　060
例 27.　丙烯装饰插画　061
例 28.　童话绘本插画　062
例 29.　点彩插画　063
例 30.　写实主义插画　064

第 3 章　DIY 设计 —————————————— 065

例 1. 水果胶带设计	066	
例 2. 植物胶带设计	067	
例 3. 植物贴纸设计	068	
例 4. 书签设计	069	
例 5. 捕梦网设计	070	
例 6. 立体贺卡设计	071	
例 7. 蝴蝶胸针设计	072	
例 8. 花盆设计	073	
例 9. 奶油胶手机壳设计	074	
例 10. 掐丝珐琅台灯设计	075	
例 11. 水晶风铃设计	076	
例 12. 糖画设计	077	
例 13. 糖霜饼干设计	078	
例 14. 藤编猫窝设计	079	
例 15. 衍纸艺术设计	080	
例 16. 纸杯定制设计	081	
例 17. 月饼礼盒设计	082	
例 18. 折纸犀牛设计	083	
例 19. 植物火漆印章设计	084	
例 20. 珠绣灯设计	085	
例 21. 桌面风扇设计	086	
例 22. 钩针童鞋设计	087	
例 23. 玉雕设计	088	
例 24. 剪纸艺术设计	089	
例 25. 定制帆布包设计	090	
例 26. 盆栽摆件设计	091	
例 27. 城市拼图设计	092	
例 28. 充电宝设计	093	
例 29. 刺绣图设计	094	
例 30. 蛋糕设计	095	
例 31. 滴胶戒指设计	096	

第 4 章　手办设计 —————————————— 097

例 1. 手办基础设计	098	
例 2. 积木手办设计	099	
例 3. 宇航员手办设计	100	
例 4. 手办三视图设计	101	
例 5. 迷你厨房手办设计	102	
例 6. 机械蜻蜓手办设计	103	
例 7. 恐龙手办设计	104	
例 8. 金属手办设计	105	
例 9. 生肖鼠手办设计	106	
例 10. 帆船手办设计	107	
例 11. 编织考拉设计	108	
例 12. 木质建筑手办设计	109	
例 13. 汉堡手办设计	110	
例 14. 存钱罐设计	111	
例 15. 汉服手办设计	112	
例 16. 桥梁模型手办设计	113	
例 17. 小天使手办设计	114	
例 18. 鲸鱼手办设计	115	
例 19. 赛博小鸟手办设计	116	
例 20. 微缩模型设计	117	
例 21. 气球小狗设计	118	
例 22. 木质猫手办设计	118	
例 23. 火箭手办设计	119	
例 24. 飞船手办设计	120	
例 25. 卫星手办设计	121	
例 26. 飞机手办设计	122	
例 27. 直升机手办设计	123	
例 28. 赛车手办设计	124	
例 29. 公交车手办设计	125	
例 30. 摩托车手办设计	126	

第 5 章　LOGO 设计 —————— 127

例 1. LOGO 基础设计　128
例 2. LOGO 个性化设计　129
例 3. LOGO 艺术化设计　130
例 4. 参考图生成 LOGO 设计　131
例 5. 字母 W 延展 LOGO 设计　132
例 6. 字母 Y 延展 LOGO 设计　133
例 7. 双鱼座 LOGO 设计　134
例 8. 巨蟹座 LOGO 设计　135
例 9. 多边形艺术风 LOGO 设计　136
例 10. 冰淇淋 LOGO 设计　137
例 11. 巴士 LOGO 设计　138
例 12. 比萨 LOGO 设计　139
例 13. 法式复古风 LOGO 设计　140
例 14. 花朵 LOGO 设计　141
例 15. 徽标 LOGO 设计　142
例 16. 男士头像 LOGO 设计　143
例 17. 鸟头 LOGO 设计　144
例 18. 人体元素 LOGO 设计　145
例 19. 美式复古 LOGO 设计　146
例 20. 剪影风格 LOGO 设计　147
例 21. 应用图标设计　148
例 22. 元素图标设计　148

第 6 章　海报主图设计 —————— 149

例 1. 立春海报主图设计　150
例 2. 芒种海报主图设计　151
例 3. 夏至海报主图设计　152
例 4. 秋分海报主图设计　153
例 5. 立冬海报主图设计　154
例 6. 端午节海报主图设计　155
例 7. 春节海报主图设计　156
例 8. 购物节海报主图设计　157
例 9. 父亲节海报主图设计　158
例 10. 母亲节海报主图设计　159
例 11. 生日海报主图设计　160
例 12. 婚庆海报主图设计　161
例 13. 国潮海报主图设计　162
例 14. 毕业季海报主图设计　163
例 15. 露营海报主图设计　164
例 16. 旅游海报主图设计　164
例 17. 饮料海报主图设计　165
例 18. 客房海报主图设计　166
例 19. 护肤品海报主图设计　167
例 20. 口红海报主图设计　168
例 21. 香薰海报主图设计　169
例 22. 珠宝海报主图设计　170
例 23. 教师节海报主图设计　171
例 24. 美食海报主图设计　172
例 25. 茶具海报主图设计　173
例 26. 音乐节海报主图设计　174
例 27. 运动会海报主图设计　175
例 28. 家具海报主图设计　176
例 29. 登山海报主图设计　177
例 30. 空镜展台海报主图设计　178
例 31. 耳机海报主图设计　179
例 32. 坤包海报主图设计　180
例 33. 香水海报主图设计　181
例 34. 汽车海报主图设计　182

第 7 章　自媒体相关设计 — 183

例 1. 参考图生成定制头像设计　184
例 2. 参考图生成情侣头像设计　185
例 3. 美妆参考设计　186
例 4. 表情包设计　187
例 5. 直播礼物特效设计　187
例 6. 灯牌设计　188
例 7. APP 开屏界面设计　189
例 8. 弹窗广告设计　190
例 9. 发布会主视觉设计　191
例 10. 数据分析 PPT 模板设计 192
例 11. 美妆直播间设计　193
例 12. 夏日氛围直播间背景设计　194
例 13. 梦幻氛围光感直播间背景设计　195
例 14. 科技感氛围直播间背景设计　196
例 15. 中国风直播间背景设计　197
例 16. 新年氛围直播间背景设计　198

第 8 章　摄影效果制作 — 199

例 1. 微距摄影效果　200
例 2. 鱼眼镜头效果　201
例 3. 广角镜头效果　202
例 4. 魔幻现实主义摄影效果　203
例 5. 极光照片效果　204
例 6. 动物类摄影效果　205
例 7. 花卉类摄影效果　206
例 8. 无人机俯拍效果　207
例 9. 红外摄影效果　208
例 10. 光绘效果　209
例 11. X 射线摄影效果　210
例 12. 双重曝光效果　211
例 13. 人物情绪摄影效果　212
例 14. 展示台摄影效果　213
例 15. 人物景别摄影效果　214
例 16. 色彩焦点照片效果　215
例 17. 散景效果　216
例 18. 动态模糊效果　217
例 19. 移轴摄影效果　218
例 20. 轮廓光摄影效果　219
例 21. 飞溅摄影效果　220
例 22. 长焦摄影效果　221
例 23. 特写摄影效果　222
例 24. 复古艺术照摄影效果　223
例 25. 封面人物摄影效果　224
例 26. 婚纱摄影效果　225
例 27. 工作照摄影效果　226
例 28. 食品素材摄影效果　227
例 29. 美食摄影效果　227
例 30. 烤箱内摄影效果　228

第 9 章　装帧及文创设计 — 229

例 1. 画册设计　230
例 2. 布艺封面设计　231
例 3. 地理书籍封面设计　232
例 4. 儿童书籍封面设计　232
例 5. 科幻书籍封面设计　233
例 6. 时尚杂志封面设计　234

例 7. 精装书封面设计	235	
例 8. 古风风格书籍封面设计	235	
例 9. 镂空封面设计	236	
例 10. 说明书排版设计	237	
例 11. 定制信封设计	237	
例 12. 线圈本设计	238	
例 13. 邀请函设计	238	
例 14. 立体贺卡设计	239	
例 15. 线装书工艺设计	240	
例 16. 折页宣传册设计	240	
例 17. 立体书设计	241	
例 18. 书脊陈列设计	242	
例 19. 明信片设计	242	
例 20. 名片设计	243	
例 21. 伴手礼设计	243	
例 22. 便签设计	244	
例 23. 特殊工艺贺卡设计	244	
例 24. 词典设计	245	
例 25. 相册排版设计	246	
例 26. 烫金工艺封面设计	247	
例 27. 定制钢笔设计	248	
例 28. 邮票设计	249	
例 29. VIP 卡设计	250	
例 30. CD 封套设计	251	
例 31. 手账版式设计	252	

第 10 章　展品设计 —— 253

例 1. 旗袍橱窗模特设计	254	
例 2. 运动服装橱窗模特设计	255	
例 3. 服装材质设计	256	
例 4. 服装风格设计	257	
例 5. 汉服元素服装设计	258	
例 6. 羽绒服服装设计	259	
例 7. 男士毛衣设计	260	
例 8. 女士连衣裙设计	261	
例 9. 男士西装设计	262	
例 10. 女士水桶包设计	263	
例 11. 男士邮差包设计	264	
例 12. 女士托特包设计	265	
例 13. 男士皮鞋设计	266	
例 14. 女士高跟鞋设计	267	
例 15. 女士高跟鞋材质设计	268	
例 16. 运动鞋设计	269	
例 17. 鞋子花纹设计	270	
例 18. 童袜设计	271	
例 19. 童鞋（运动鞋）设计	272	
例 20. 宠物衣服设计	273	
例 21. 家纺设计	274	
例 22. 地毯花纹设计	275	
例 23. 珠宝设计	276	
例 24. 模特珠宝展示设计	277	
例 25. 运动手表设计	278	
例 26. 女士手表设计	279	
例 27. 户外充电器设计	280	
例 28. 户外登山包设计	281	
例 29. 户外帐篷设计	282	
例 30. 餐具设计	283	
例 31. 酸奶包装盒设计	284	
例 32. 产品系列化设计	285	
例 33. 儿童闹钟设计	286	
例 34. 儿童车设计	287	
例 35. 唱片机设计	288	
例 36. 耳机设计	289	
例 37. 猫窝设计	290	
例 38. 换鞋凳设计	291	
例 39. 化妆品套装设计	292	
例 40. 咖啡机设计	293	
例 41. 饼干包装铁盒设计	294	
例 42. 蓝牙音箱设计	295	

例 43. 破壁机设计	296	例 46. 适老化轮椅设计	299
例 44. 剃须刀设计	297	例 47. 展厅设计	300
例 45. 吸尘器设计	298		

第 11 章　室内设计 —————— 301

例 1. 厨房设计	302	例 17. 书柜设计	318
例 2. 儿童书房设计	303	例 18. 台灯设计	319
例 3. 衣帽间设计	304	例 19. 衣柜设计	320
例 4. 卧室设计	305	例 20. 瓷砖设计	321
例 5. 大门设计	306	例 21. 阳台设计	322
例 6. 儿童房设计	307	例 22. 服装店铺设计	323
例 7. 茶室设计	308	例 23. 咖啡店铺设计	324
例 8. 中式餐厅设计	309	例 24. 主题快闪店设计	325
例 9. 绿植角设计	310	例 25. 移动咖啡车设计	326
例 10. 展示大厅设计	311	例 26. 房车内部设计	327
例 11. 家庭影院设计	312	例 27. 书店设计	328
例 12. 卫生间设计	313	例 28. 珠宝店设计	329
例 13. 铁艺床设计	314	例 29. 博物馆设计	330
例 14. 鞋柜设计	315	例 30. 新中式茶室设计	331
例 15. 沙发设计	316	例 31. 水下餐厅设计	332
例 16. 书桌设计	317		

1.1 Midjourney 的安装与订阅

Midjourney 是一款搭载在 Discord（一款便于在游戏内即时通信的聊天软件）服务器上的 AI 绘画工具，由 Midjourney 研究实验室开发。它可以根据文本或图像生成图像，读者可通过向 Midjourney 的机器人发送命令来创作出图像作品。

Midjourney 非常适合初学者使用，简单易学，生成的图像质量高，速度快。如果你想学习 AI 绘画，Midjourney 是一个不错的选择，只要跟着本书一步一步学习，就可以快速入门。

1.1.1 注册

由于 Midjourney 是搭载在 Discord 上运行的，因此，如果读者想要体验 Midjourney，就需要先注册 Discord 账号。

步骤① 打开 Midjourney 官网，如图 1.1-1 所示。

图 1.1-1

步骤② 单击右下角的"Join the Beta"按钮，进入 Discord 的注册页面，如图 1.1-2 所示。
步骤③ 在"昵称"输入框中输入自己的昵称，单击"继续"按钮，如图 1.1-3 所示。

图 1.1-2

图 1.1-3

步骤④ 在弹出的确认对话框中勾选"我是人类"复选框，完成测试，如图 1.1-4 所示。
步骤⑤ 进入确认年龄的界面，输入出生日期后单击"下一步"按钮，如图 1.1-5 所示。

第 1 章 Midjourney 入门

图 1.1-4

图 1.1-5

> **Tips**
> 在这里需要注意：年龄一定要填写为 18 岁以上，否则会被驳回使用请求，并且后续的关联账号也没有办法重新申请。

步骤⑥ 在弹出的对话框中填写注册邮箱和密码，然后单击"认证账号"按钮，如图 1.1-6 所示。

步骤⑦ 进入注册的邮箱，确认注册后，Discord 账号就注册完成了，如图 1.1-7 所示。

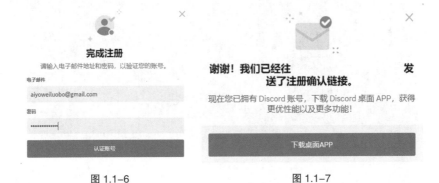

图 1.1-6　　　　　　　　　　图 1.1-7

1.1.2　登录

步骤① 在浏览器搜索栏中输入 Discord 官网地址，进入网站主页，单击"在您的浏览器中打开 Discord"按钮，如图 1.1-8 所示。

图 1.1-8

步骤② 进入登录页面,输入注册的邮箱和密码,单击"登录"按钮即可登录使用,如图 1.1-9 所示。

图 1.1-9

1.1.3 安装

用户可以直接在网站上登录使用 Discord,也可以将 Discord 客户端下载到计算机,通过客户端使用。

步骤① 在浏览器搜索栏中输入 Discord 官网地址,进入网站主页,单击"Windows 版下载"按钮,如图 1.1-10 所示。

图 1.1-10

步骤② 在网站页面右上角弹出的对话框中单击"另存为"按钮,如图 1.1-11 所示。

图 1.1-11

步骤③ 在弹出的"另存为"对话框中选择存储位置,单击"保存"按钮,如图 1.1-12 所示。

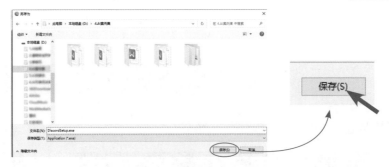

图 1.1-12

步骤④ 下载完成后，单击网站页面右上角的"打开文件"链接，即可开始安装，如图 1.1-13 所示。

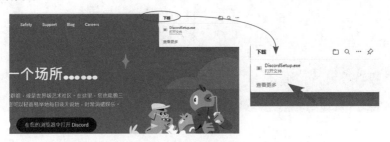

图 1.1-13

步骤⑤ 安装完成后即可进入登录界面，登录后便可开始使用，如图 1.1-14 所示。

图 1.1-14

步骤⑥ 如果想要能够直接在电脑桌面单击图标使用 Discord，可进入保存 Discord 的文件夹中，选择 Discord 的图标并右击，在弹出的快捷菜单中选择"创建快捷方式"命令即可，如图 1.1-15 所示。

图 1.1-15

1.1.4 创建个人服务器

进入 Midjourney 主界面后，在公共服务器内就可以使用绘画功能了。但是，因为公共服务器使用人数较多，用户的绘画内容容易被淹没，所以建立自己的个人服务器就十分有必要。

① 创建个人服务器

下面详细讲解如何创建个人服务器。

步骤① 单击主界面左侧的 ➕ 按钮，如图 1.1-16 所示。

图 1.1-16

步骤② 在弹出的对话框中单击"亲自创建"按钮，如图 1.1-17 所示。

步骤③ 在弹出的对话框中单击"仅供我和我的朋友使用"按钮，如图 1.1-18 所示。

图 1.1-17

图 1.1-18

步骤④ 单击上传头像 按钮，在"服务器名称"输入框中输入自己的服务器名称，再单击"创建"按钮，如图 1.1-19 所示。

图 1.1-19

步骤⑤ 当显示欢迎提示时，说明个人服务器已经创建成功，如图 1.1-20 所示。

图 1.1-20

②添加 Midjourney 机器人

创建个人服务器相当于创建了一个可以独立使用的房间，但如果没有 Midjourney 机器人，用户还是无法绘画，接下来，需要添加 Midjourney 机器人进入用户创建的房间提供绘画服务。

步骤① 单击 Midjourney 帆船图标，进入 Midjourney 公共服务器，如图 1.1-21 所示。

图 1.1-21

步骤② 进入公共服务器后，单击左侧菜单栏中的"NEWCOMER ROOMES（新人房间）"，如图 1.1-22 所示。

步骤③ 单击界面中的 Midjourney 机器人，如图 1.1-23 所示。

图 1.1-22

图 1.1-23

步骤④ 单击"添加至服务器"按钮，如图 1.1-24 所示。

步骤⑤ 在"选择一个服务器"下拉列表中选择自己创建的服务器，并单击下方的"继续"按钮，如图 1.1-25 和图 1.1-26 所示。

图 1.1-24

图 1.1-25

图 1.1-26

步骤⑥ 单击"授权"按钮并完成相应的测试，就可以开始使用，如图 1.1-27 所示。

图 1.1-27

1.1.5 订阅服务

Midjourney 的绘画服务需要收取一定的费用，所以用户在绘画之前需要进行订阅。

进入自己创建的服务器后，在下方的输入框中输入 /subscribe 命令，打开网站后即可选择适合自己的套餐模式并购买，见表 1.1-1。

表 1.1-1

	标准版	高级版	专业版
月付价	10 美元 / 月	30 美元 / 月	60 美元 / 月
年付价	96 美元 / 年	288 美元 / 年	576 美元 / 年
快速（Fast）模式	200 张 / 月	15 小时 / 月	30 小时 / 月
轻松（Relax）模式	不支持	无限制	无限制
社区画廊	支持	支持	支持
并发作业数	3 个	3 个	3 个
商业模式	支持	支持	支持

Midjourney 的订阅套餐有以下几种：

（1）快速（Fast）模式：高优先级的出图模式，出图速度更快，一个小时大约可生成 60 张图像（非准确数字）。

（2）轻松（Relax）模式：排队出图模式，视服务器拥挤程度而定。

（3）社区画廊：官方提供的高质量素材集合，并且可以通过图像查看原始命令。

（4）并发作业数：服务器最多可同时处理描述词的组数，若组数过多，超出部分会显示"排队中"，待处理完前置任务后，服务器才会继续处理。

（5）商业模式：可将图像应用于任何商业化场景。

1.2 用文字生成图像

Midjourney 常用的生成图像方式有 3 种：用文字生成图像、用图像生成图像和通过混合图像生成图像。而用文字生成图像是比较常用的方式。

在 Midjourney 中，用户只需输入绘画内容提示词，Midjourney 即可根据提示词生成图像。

步骤① 在界面下方的输入框中输入"/"，然后在弹出的列表中选择 imagine 命令，如图 1.2-1 所示。

图 1.2-1

步骤② 在 prompt 输入框中输入想要绘画内容的提示词。例如，这里想画一个穿蓝色衬衫的女孩，就可以直接输入：a girl in a blue shirt，并按 Enter 键发送命令，如图 1.2-2 所示。

图 1.2-2

步骤③ 等待片刻，即可生成图像，如图 1.2-3 所示。

图 1.2-3

1.3 出图界面常用按钮介绍

Midjourney 可以一次输出 4 张图像，下面以本次输出的图像为例，详细讲解出图界面中常用按钮的作用。

1.3.1 U1–U4 按钮

U1–U4 按钮是指图像从左至右、从上到下的顺序，单击相应的按钮，就能得到相应序号图像单独输出的图像，如图 1.3-1 所示。

图 1.3-1

例如，单击 U2 按钮，则会单独输出右上角图像，如图 1.3-2 所示。

图 1.3-2

1.3.2 🔄 按钮的作用

在 U4 按钮右边的 🔄 按钮，其功能是重新生成 4 张图像。

步骤① 单击 🔄 按钮。

步骤② 在弹出的对话框中，如果直接单击"提交"按钮，则会重新生成 4 张图像，如图 1.3-3 所示。

步骤③ 如果需要调整提示词，还可以对提示词进行修改，如这里将 girl 更改为 boy，如图 1.3-4 所示。单击 提交 按钮。

图 1.3-3

图 1.3-4

步骤④ 得到修改提示词后重新生成的图像，如图 1.3-5 所示。

图 1.3-5

1.3.3 V1-V4 按钮的作用

V1-V4 按钮同上面 U1-U4 按钮一样，也是代表了从左至右、从上至下的图像顺序，不同的是，V 代表了图像的风格，可以根据所选按钮，重新生成与按钮对应图像风格相似的 4 张图像。例如，在 V4 风格的基础上再生成 4 张图像。

步骤① 单击 V4 按钮，如图 1.3-6 所示。

步骤② 在弹出的对话框中单击"提交"按钮，如图 1.3-7 所示。

图 1.3-6

图 1.3-7

步骤③ 生成与 V4 对应的右下角图像风格相似的 4 张新图像,如图 1.3-8 所示。

图 1.3-8

如果想要在该风格基础上对绘画内容进行修改,可单击 V4 按钮,在弹出的对话框中对文本进行修改。例如,在 V4 风格的基础上生成 4 张主体为一个男孩的图像。

步骤① 单击 V4 按钮。

步骤② 在弹出的窗口输入框中将"A girl in a blue shirt"修改为"A handsome man, 28 years old(一个帅气的男孩,28 岁)",再单击"提交"按钮,如图 1.3-9 所示。

图 1.3-9

步骤③ 生成 4 张与 V4 按钮对应图像相似风格的男孩图像,如图 1.3-10 所示。

图 1.3-10

> **Tips**
> 由于我们是在女孩图像的基础上进行修改，因此生成的图像可能仍会有部分趋近女孩形象，这个时候，我们可以选择重复操作 V 系列按钮，直到得到想要的图像效果。

1.4 打开和保存图像

在 Midjourney 中生成的图像，用户可以在浏览器中打开进行查看，如果需要，可以将图像保存到本地磁盘进行使用。

1.4.1 在浏览器中打开图像

得到一张单独输出的图像后，如果想观察更多细节，可以先单击图像，然后单击图像左下角的"在浏览器中打开"链接，就能更清楚地观察生成图像的细节，如图 1.4-1 所示。

1.4.2 保存图像

在 Midjourney 中生成理想的图像后，用户就可以将图像保存下来，用于工作和生活中。

在浏览器中打开图像后右击，在弹出的快捷菜单中选择"将图像另存为"命令，在弹出的"另存为"对话框中选择保存的位置，在"文件名"输入框中输入文件名称，然后单击"保存"按钮，如图 1.4-2 和图 1.4-3 所示。

图 1.4-1

图 1.4-2

图 1.4-3

1.5 重绘按钮和 Zoom 按钮

在选择输出单独的图像后，图像下面会跟随 4 排新的功能按钮，它们的作用是什么呢？下面就来依次了解一下。

1.5.1 重绘按钮

图像下面第 1 排的后 3 个按钮主要用于图像重绘,如图 1.5-1 所示。

图 1.5-1

(1) Vary (Strong)[变化 (强)]:在原图基础上重新生成 4 张新的图像,风格相同,但在细节上会明显不同,如发饰、脸部细节和背景等,如图 1.5-2 所示。

(2) Vary (Subtle)[变化 (微妙)]:在原图基础上重新生成 4 张新的图像,但在细节上有微妙的不同,如眼神、发丝等,如图 1.5-3 所示。

图 1.5-2   图 1.5-3

(3) Vary (Region)[变化 (局部)]:在原图基础上可以重新框选部分区域,并对该区域内容进行重新生成。

这里以将衣服更换为"Black clothes"(黑色衣服)为例进行讲解。

步骤① 在生成的图像中选择一张需要进行局部调整的图像,如图 1.5-4 所示。

图 1.5-4

步骤② 在单独输出的高清图像中单击局部重绘 Vary(Region)按钮,如图 1.5-5 所示。

图 1.5-5

步骤③ 在弹出的窗口中选择左下角的套索工具 ,然后选择需要重新生成的区域,如图 1.5-6 所示。

图 1.5-6

步骤④ 在局部重绘窗口下方选择需要更改的提示词,如图 1.5-7 所示。

图 1.5-7

步骤⑤ 将提示词更改为"Black clothes",如图 1.5-8 所示。

图 1.5-8

步骤⑥ 单击输入框右侧的 ▶ 按钮,发送提示词,生成局部重绘的图像,如图 1.5-9 所示。

图 1.5-9

除了套索工具,使用框选工具 同样能达到不错的效果。二者的区别是,套索工具适合不规则区域的框选,框选工具适合规则区域的框选。这里以用框选工具将人物眼睛由睁眼生成闭眼为例来讲解。

步骤① 在单独输出的高清图像中单击局部重绘 Vary(Region)按钮,如图 1.5-10 所示。

步骤② 在弹出的窗口中选择左下角的框选工具,然后选择需要重新生成的区域,此处框选眼睛处,如图 1.5-11 所示。

图 1.5-10

图 1.5-11

步骤③ 在局部重绘窗口下方选择需要更改的提示词,如图 1.5-12 所示。

图 1.5-12

步骤④ 将提示词更改为"eyes closed"(闭着眼睛),如图 1.5-13 所示。

图 1.5-13

步骤⑤ 单击输入框右侧的 ▶ 按钮,发送提示词,生成局部重绘后的图像如图 1.5-14 所示。

图 1.5-14

1.5.2 Zoom 按钮的使用

在实际应用中,用户可能对 Midjourney 生成的图像比较满意,但在将该图像制作成海报时,有时需要更多的画面空间来放文字,这时就可以使用 Zoom 按钮对图像的主体进行缩小。

(1) Zoom Out 2x:将画面推远 2 倍,如图 1.5-15 所示。

图 1.5-15

（2）Zoom Out 1.5x：将画面推远 1.5 倍，如图 1.5-16 所示。

图 1.5-16

（3）Custom Zoom：自定义缩放大小。

1.5.3 无损放大功能

在 Midjourney 生成图像后，如果需要更高的图像分辨率，就可以使用 Upscale 按钮对图像进行无损放大。

（1）Upscale (2x)[高档 (2x)]：生成图像，分辨率无损放大 2 倍，如图 1.5-17 和图 1.5-18 所示。

（2）Upscale (4x)[高档 (4x)]：生成图像，分辨率无损放大 4 倍，如图 1.5-19 所示。

原图

分辨率：1024×1024

图 1.5-17

2 倍

分辨率：2048×2048

图 1.5-18

4 倍

分辨率：7096×7096

图 1.5-19

1.5.4 Web 网页按钮

在 Midjourney 中，所有创作的图像都是同步到 Midjourney 官网中的，用户想在官网中查看图像时，就可以通过 Web 网页按钮一键到达。

Web 网页按钮：在官网内浏览图像，单击此按钮后，系统自动跳转到 Midjourney 官网。

1.6 设置面板的基本功能

为了更好地使用 Midjourney，用户需要对更多的命令进行熟悉和了解。

1.6.1 常用基础命令的功能

在输入框中输入"/"，Midjourney 会自动弹出一系列基础命令，下面简单说明不同基础命令的基本功能，如图 1.6-1 所示。

图 1.6-1

（1）/imagine（想象）：用文本生成图像。（具体操作方法详见 1.2 节）

（2）/describe（描述）：上传图像后，Midjourney 可以根据图像生成 4 组对应的提示词。（具体操作方法详见 1.10 节）

（3）/blend（混合）：上传 2～5 张图像，AI 自动生成融合上传图像要素的新图像。（具体操作方法详见 1.9 节）

（4）/settings（设置）：打开 Midjourney 的出图设置面板，对各种参数进行设置。（具体操作方法详见 1.6.2 小节）

（5）/subscribe（订阅）：前往 Midjourney 官网，打开订阅页面。（具体操作方法详见 1.1.5 小节）

（6）/ask（提问）：向 Midjourney 机器人提问，寻求使用帮助。如果用户想知道如何在 Midjourney 中删除已生成或上传的图像，就可以用该命令提问。

下面介绍 /ask 命令的使用方法。

步骤① 在输入框中输入 /ask 命令，按 Enter 键，即可出现一个 question 的输入框，如图 1.6-2 所示。

步骤② 在输入框中输入问题"How do I delete an image?"，如图 1.6-3 所示。

第 1 章 Midjourney 入门

图 1.6–2

图 1.6–3

步骤③　输入完成后，按 Enter 键，Midjourney 会回复用户的提问，如图 1.6-4 所示。

图 1.6–4

1.6.2 设置功能面板

了解一些基础命令后，选择 /settings 命令，按 Enter 键即可打开设置功能面板，然后在该面板中对出图的各项参数进行设置。下面进行详细介绍，如图 1.6-5 所示。

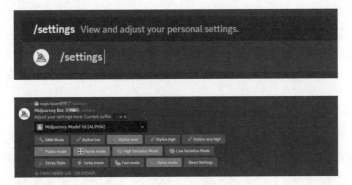
图 1.6–5

在"Adjust your settings here. Current suffix：" (在这里调整您的设置，当前后缀:) 下的下拉列表中可以选择 Midjourney 出图的版本。默认为 Midjourney Model V6 [ALPHA]，如图 1.6-6 所示。

图 1.6–6

单击"Midjourney Model V6 [ALPHA]"后的 ⌄ 按钮，在弹出的下拉列表中可以选择不同的版本。

Midjourney 的版本大致可以分为两种类型：

一种是 V 系列的版本，该版本的出图效果偏向于写实，版本越高，艺术感越强。

另一种是 Niji 类型的版本，画风更偏向于动漫风格，当需要创建动漫风格的图像时，就可以选择 Niji 模式。

> **Tips**
>
> 除了可以通过设置面板来选择生成图像的版本外，也可以通过直接在提示词后添加参数的方式来选择，如果要选择 Niji 6 版本，则在提示词后输入：空格 + "--niji" + 空格 + "6"。例如：A handsome man, 28 years old, --niji 6。

除了版本选择外，其他的设置按钮大致可以分为以下几类：

1. 模式

RAW Mode：回归模式，即回归到写实的模式，单击这个按钮，后续在这个模式下生成的图像会更接近真实质感。

2. 风格化等级

风格化越高，AI 介入度越高，生成的图像艺术化越高；风格化越低，AI 自由发挥的空间越低，生成的图像越接近所描述的提示词。一般来说，系统默认值为 100。

（1）Stylize low（风格化低）：相当于在提示词中输入风格化参数"--s 50"的效果，如图 1.6-7 所示。

（2）Stylize med（风格化中等）：相当于在提示词中输入风格化参数"--s 100"的效果，如图 1.6-8 所示。

图 1.6-7

图 1.6-8

（3）Stylize high（风格化高）：相当于在提示词中输入风格化参数"--s 350"的效果，如图 1.6-9 所示。

（4）Stylize very high（风格化非常高）：相当于在提示词中输入风格化参数"--s 750"的效果，如图 1.6-10 所示。

图 1.6-9

图 1.6-10

> **Tips**
>
> 用户可以根据需要将风格化参数值设置在 0 ~ 1000 之间。同一关键词内,风格化 "--s 10" 和 "--s 1000" 的区别如图 1.6-11 和图 1.6-12 所示。
>
>
> 图 1.6-11
>
>
> 图 1.6-12

3. 图像模式

(1) Public mode:公开模式,任何用户都能看到生成的图像,默认是打开的,只有订阅套餐在 60 美元 / 月等级以上的用户才能取消。

(2) Remix mode:混合模式,在生成图像之后,可以修改提示词再继续生成。

(3) High Variation Mode:高变化模式,在生成的一组图像中,4 张图中每一张的差异十分明显。

(4) Low Variation Mode:低变化模式,在生成的一组图像中,4 张图风格类似,差异较低。

4. 出图速度

(1) Turbo mode:涡轮模式,出图速度超快,但需要订阅套餐达到 120 美元 / 月时才能使用。

(2) Fast mode:快速出图模式,正常订阅都能使用,但有时间限制,不同的订阅套餐,所需的时间也不同。

（3）Relax mode：放松模式，速度最慢，但是不限出图量。

5. 重置

Reset Settings：重置按钮，单击该按钮后，所有设置回归到初始模式。

1.7 常见参数

在 Midjourney 中，除了可以在设置面板中对参数进行设置外，还有一种比较简便的方式，就是在提示词后直接输入想设置的参数。

1.7.1 参数的输入方法

在 Midjourney 中，参数往往是在提示词的最后输入，其格式为：空格 +--+ 参数名称 + 空格 + 参数数值。例如，Midjourney 默认生成的图像比例为 1:1，当需要生成一张比例为 3:4 的图像时，就可以在提示词的最后输入"--ar 3:4"。这样生成图像的比例就是 3:4，如图 1.7-1 所示。

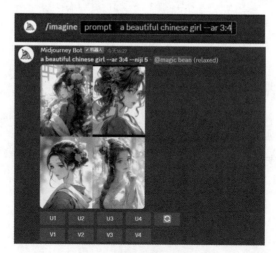

图 1.7-1

当有多个参数时，参数之间也需要用空格隔开。其格式为：空格 + 参数 1 名称 + 空格 + 参数 1 参数值 + 空格 + 参数 2 名称 + 空格 + 参数 2 参数值……这里需要注意的是，参数越靠前，优先级越高。因此，在输入参数时，如果涉及多个参数，则需要考虑参数的先后顺序。

1.7.2 常见参数一览

在使用 Midjourney 生成图像时，有一些参数较为常用，其参数名称如下：

--aspect 或 --ar：设置宽高比。

--quality 或 --q：设置图像质量（画质），设置区间为 0.25～5，数值越高，画面细节越丰富。
--version 或 --v：切换算法模型，如 V5.2、V6 等。
--niji：使用 Niji 模型（动漫风）。
--chaos 或 --c：设置创意程度，数值为 0～100，数值越大，同一批次 4 张图像的差异越大。
--stylize 或 --s：设置艺术性。
--seed/--sameseed：获取 seed（种子）值。
--no：设置不要出现的元素。
--stop：暂停生成进度，设置数值为 0～100，设置为多少，则生成进度会暂停到该数值内，如设置为"--stop 50"，则生成进度会暂停到 50%。
--tile：生成无缝贴图。
--iw：设置参考图片（俗称垫图）的权重比，数值越低，对原图的参考越小；数值越高，对原图的参考越高。（具体操作方法详见 1.8 节）
--creative 与 --test/--testp：使用测试算法模型（增加创意）。
--uplight：在放大图像时添加少量细节纹理。
--upbeta：在放大图像时不添加细节纹理。
--upanime：在放大图像时增加动画插画风格。
--hd：生成高清图。
--video：生成渲染过程的演示视频（在 V4 和 V5 版本中无法使用）。
--repeat：重复生成图像。

1.8 图生图的使用方法

在 Midjourney 中，除了可以通过输入提示词生成图像外，还可以通过上传图片来生成图像。

通过图生图的方式生成图像，就是垫图，可以理解为用户为 Midjourney 绘画机器人提供一张参考图，并告诉它参考这张图进行生成。需要注意的是，Midjourney 更注重模仿图像的风格。

下面以生成一张像自己的卡通头像图为例来讲解图生图的使用方法。

步骤① 双击输入框左侧的加号，在弹出的菜单栏中选择"上传文件"选项，如图 1.8-1 所示。
步骤② 在弹出的对话框中选择需要上传的图片，单击右下角的"打开"按钮，如图 1.8-2 所示。

图 1.8-1

图 1.8-2

步骤③ 当图片出现在操作界面中时，按 Enter 键，对图片进行上传，如图 1.8-3 所示。

步骤④ 等图片上传后右击，在弹出的快捷菜单中选择"复制链接"命令，如图 1.8-4 所示。

图 1.8-3　　　　　　　　　　　　图 1.8-4

Tips

注意：需要选择"复制链接"命令而不是"复制消息链接"命令。

步骤⑤ 复制链接后，按照正常的文生图流程，在输入框中输入"/"，选择 imagine 命令，然后粘贴刚才复制的图像链接，如图 1.8-5 所示。

图 1.8-5

步骤⑥ 输入空格，区分开图像链接和绘画提示词内容，再输入想要生成图像的提示词。例如，这里希望 Midjourney 参考该图片，生成一张年轻女孩的皮克斯卡通风格头像图，就输入：Portrait of young girl, Pixar style, cute cartoon, C4D rendering, --iw 1.5 --v 5.2（年轻女孩肖像，皮克斯风格，卡通可爱，C4D 渲染，--iw 1.5 --v 5.2），如图 1.8-6 所示。

图 1.8-6

步骤⑦ 按 Enter 键，即可看到参考上传图片后生成的皮克斯卡通风格的头像图，如图 1.8-7 所示。

第 1 章 Midjourney 入门

图 1.8-7

> **Tips**
>
> 下面重新看这段提示词，总结一下垫图的关键要素。
>
> https://s.mj.run/O0tm3dhsqEI Portrait of young girl, Pixar style, cute cartoon, C4D rendering, --iw 1.5 --v 5.2
>
> 　　通过这张图片可以看到，这段提示词主要分为三个部分：参考图链接 + 内容描述 + 后缀参数，每一个部分用空格符号区分开，这就是垫图的公式了。
>
> 　　两个后缀词的意思分别是：
>
> iw：与原图的相似权重，一般设置在 0.5～2 之间，数值越高，相似度越高。
>
> V 5.2：绘图模型，如果想要更换为二次元或者动漫风格，可以更改为 Niji 模型。

1.9　多张图片混合生成图像的方法

　　多张图片混合生成图像，即提供几张图片，让 Midjourney 参考这几张图片的元素，重新生成新的图像。到目前为止，Midjourney 支持上传 2～5 张图片。这里以两张图片混合为例进行讲解。

步骤①　在输入框中输入 "/"，然后选择 blend 命令，如图 1.9-1 所示。

图 1.9-1

步骤② 在弹出的对话框中单击上方的 image1 方框，如图 1.9-2 所示。

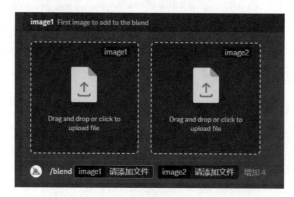

图 1.9-2

步骤③ 在弹出的对话框中选择第 1 张需要上传的图片，单击"打开"按钮，如图 1.9-3 所示。

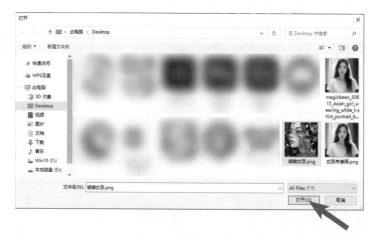

图 1.9-3

步骤④ 单击 image2 方框，按照上一步的方法继续上传第 2 张图片，如图 1.9-4 所示。

图 1.9-4

步骤⑤ 上传完成后，按 Enter 键，即可看到两张图片混合后生成的图像了，如图 1.9-5 所示。

第1章 Midjourney 入门

图 1.9-5

> **Tips**
> 　　如果想混合更多的图片，可在 image2 方框后单击，在弹出的列表中依次选择 image3、image4、image5 即可。dimensions 命令用于指定混合后的图像尺寸，可以选择 Portrait（竖图）、Square（正方形图）、Landscape（横图），如果不选择，则默认 1:1 正方形图。

1.10 反推提示词（/describe 命令的使用）

　　当用户手上有一张图片，希望能生成与此图片相似风格的图像，但不知道该如何描述时，就可以使用 Midjourney 的 describe 命令，describe 命令将会提供 4 组提示词来供参考。

　　步骤① 在输入框中输入"/"，然后选择 describe 命令，如图 1.10-1 所示。
　　步骤② 在弹出的对话框中单击 image 方框，如图 1.10-2 所示。
　　步骤③ 在弹出的对话框中选择需要上传的图片，单击右下角的"打开"按钮，如图 1.10-3 所示。
　　步骤④ 按 Enter 键，上传图片，如图 1.10-4 所示。
　　步骤⑤ Midjourney 显示出 4 组提示词，如图 1.10-5 和图 1.10-6 所示。
　　步骤⑥ 每组提示词都能重新生成 4 张图片，图像下方的蓝底白字 1～4 的数字方框即对应了 4 组提示词。例如，单击方框，在弹出的窗口中可以修改提示词，修改完成后，可单击"提交"按钮进行提交，如图 1.10-7 所示。

步骤⑦ 此时将显示根据第 1 组提示词生成的新图像,如图 1.10-8 所示。

图 1.10-1

图 1.10-2

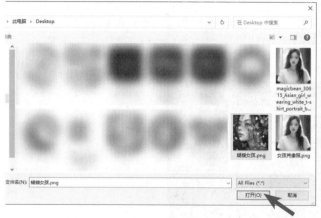

图 1.10-3

图 1.10-4

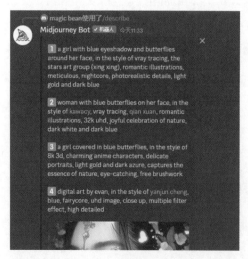

图 1.10-5

图 1.10-6

第1章 Midjourney 入门

图 1.10-7　　　　　　　　　　图 1.10-8

步骤⑧　如果对本次生成的 4 组提示词效果都不满意，可以返回步骤⑥，单击右下角的 🔄 按钮，重新生成新的 4 组提示词，如图 1.10-9 所示。

图 1.10-9

1.11 万能绘画公式的使用

在使用 Midjourney 生成图像时，特别是作为初学者，可能不知道该如何去描述想要的画面，这时推荐使用万能绘画公式，即绘画主体 + 场景 + 风格 + 画质 + 基础设置。

1.11.1 万能绘画公式的使用

如果想要画一束放在桌子上面的鲜花，就可以按万能绘画公式的结构来输入提示词。

031

绘画主体：A bunch of flowers on the table, in front of the window, wooden table, glass vase, filled with various flowers（桌上一束鲜花，窗前，木桌，玻璃花瓶，盛满各种鲜花）

场景：Morning light, asymmetrical composition（晨光，不对称构图）

风格：realistic photography style（写实摄影风格）

画质：Cinema 4D（电影 4D）

基础设置：--ar 4:3 --v 5.2（画面比例 4:3　版本 V5.2）

生成效果如图 1.11-1 所示。

图 1.11-1

1.11.2　替换万能绘画公式中的要素

在本书的案例中，以万能绘画公式的结构给出了 300 多个例子的提示词，用户可以根据需求任意进行搭配，以生成自己的绘画公式。这里以 1.11.1 小节中生成的图像重新生成线稿草图为例进行讲解。

步骤① 对提示词进行分析，"线稿草图"是用户希望生成图像呈现的风格，所以将提示词中的"Realistic photography style"（写实摄影风格）替换为"Line art sketch style"（线稿草图风格）。

步骤② 风格替换后，要根据风格选择更适合呈现的画质渲染器，所以将画质内的"Cinema 4D"（电影 4D）渲染器更改为"SketchUP"（草图大师）渲染器。

步骤③ 为了让图像呈现更加平面的效果，可以将版本由"V5.2"更改为"Niji"。

下面看一下更改后的提示词：

绘画主体：A bunch of flowers on the table, in front of the window, wooden table, glass vase, filled with various flowers（桌上一束鲜花，窗前，木桌，玻璃花瓶，盛满各种鲜花）

场景：Morning light, asymmetrical composition（晨光，不对称构图）

风格：Line art sketch style（线稿草图风格）

画质：SketchUP（草图大师）

基础设置：--ar 4:3 --niji 5（画面比例 4:3　版本 Niji5）

替换风格后生成的图像效果如图 1.11-2 所示。

图 1.11-2

Tips

如果希望生成的图像没有颜色，可以去掉光线的提示词 morning light（晨光），以及将风格更改为 Black and white line art sketch style（黑白线稿草图风格）。

1.12 文字的使用

在Midjourney V6版本中可以识别文字,但是目前只能识别出英文,在提示词中将文本内容放在""中即可。例如,输入"A child holds a wooden sign that says 'Hello'--v 6.0"(一个小孩举着写有"你好"的木牌),按Enter键,即可看到生成图像中的牌子上写着Hello,如图1.12所示。

图 1.12

第 2 章
插画绘制

例1. 插画基础

绘画主体

There is a kitten on the windowsill, the sun shone on its fur, in front of the window, glass vase, filled with various flowers
窗台上有只小猫，光洒在它的毛发上，窗前，玻璃花瓶，盛满各种鲜花

场景	风格	画质	基础设置
Morning light, asymmetrical composition 晨光，不对称构图	Illustrations 插画	HD 高清	--ar 3:4 --v 5.2 画面比例 3:4 版本 V5.2

绘画主体＋场景＋风格＋画质＋基础设置。（约定：蓝色对应绘画主体，红色对应场景，绿色对应风格，黄色对应画质，紫色对应基础设置）

例 2. 线稿插画

绘画主体

A castle on top of a mountain, clouds, staircase
坐落在山上的城堡，云朵，楼梯

场景

Asymmetrical composition
不对称构图

风格

Line draft, coloring book style, black and white lines, manuscript, smooth lines
线稿，涂色画册风格，黑白线条，手稿，线条流畅

画质

HD
高清

基础设置

--ar 3:4 --niji 5
画面比例 3:4　版本 Niji5

为了让出图效果更稳定，避免出现三张彩色图，一张线稿图的现象，可以把风格的关键词放在最前面，这样权重更大，出图更稳定。

例 3. 油画风插画

绘画主体

A girl sitting on the windowsill, long white dress, strappy sandals, long brown hair, smiling face

坐在窗台边的少女，白色长裙，绑带凉鞋，棕色长发，微笑的脸庞

场景

Window, hanging clusters of flowers, lush scenery, light white, soft lights, asymmetrical composition, knee shot, depth of field

窗台，垂下来的花簇，郁郁葱葱的景色，浅白，柔光，不对称构图，膝盖镜头，景深

风格	画质	基础设置
Oil painting style, in the style of Etienne Adolphe Piot, classical elegance, plein air painting, lots of strokes, palette knife painting 油画风格，埃蒂安·阿道夫·皮奥特风格，古典优雅，外光派画法，加强笔触，调色刀风格	HD 高清	--v 5.2 版本 V5.2

例 4. 国画风插画

绘画主体

Chinese mountains and waters, scattered and towering peaks, a stone bridge, meandering river, mountains loom in the distance, mist, flocks of birds fly over the river, floating cloud
中国山水，错落有致的山峰，一座石桥，蜿蜒的河流，远处的山若隐若现，雾气，河上有鸟群，飘浮的云

场景
A serene landscape
宁静的风景

风格
Chinese wash painting
中国水墨画

画质
16K quality
16K 画质

基础设置
--ar 2:1 --v 5.2
画面比例 2:1　版本 V5.2

例5. 敦煌风插画

绘画主体

A Chinese girl in Hanfu, dancing in Dunhuang, long black hair, wearing gold and red clothes, the sun shone on her, exquisite facial features, ultra-detailed, exquisite headwear

穿着汉服的中国女孩，跳着敦煌舞蹈，长长的黑发，穿着金红相间的衣服，阳光照在她身上，精致的五官，超细节，精致的头饰

场景

The background is ancient Chinese architecture, surrounded by colored streamers, portrait, dynamic angle, natural light, color complementation

背景是中国古代建筑，周围有彩色的飘带，人物肖像，动态角度，自然光，色彩互补

风格

Dunhuang style
敦煌风

基础设置

--niji 5
版本 Niji5

画质

8K quality
8K 画质

例 6. 人物设计插画

绘画主体

The girl who drives the race car, black hair, holding the red racing helmet, in racing gear

开赛车的女孩,黑发,抱着红色的赛车头盔,穿着赛车服

场景	风格	画质	基础设置
Clean background, full body	Character sheet	HD	--niji 5
干净的背景,全身	角色设计	高清	版本 Niji5

例 7. 木刻版画

绘画主体
A LOGO of a deer, vintage imagery
一只鹿的标志，复古的图像

场景
High contrast, dynamic line work, detailed crosshatching, organic contours
高对比度，动态的线条工作，非常细节的交叉排线，有机轮廓

风格
Woodcut, in the style of wood engraving, monochromatic compositions, hatching, abstract minimalism, letterpress, etching
木刻版画，木雕，单色构图，影线画，抽象的极简主义风格，凸版印刷，蚀刻

画质
HD
高清

基础设置
--ar 3:4 --v 5.2
画面比例 3:4　版本 V5.2

例 8. 彩色玻璃插画

绘画主体

The lighthouse by the sea, hyper detailed
海边的灯塔,超详细

风格

Stained-glass, in style by Louis Comfort Tiffany, intricate stained glass design
彩色玻璃,路易斯·康福特·蒂芙尼风格,
复杂的彩色玻璃设计

画质	基础设置
8K quality	--ar 3:4 --v 5.2
8K 画质	画面比例 3:4　版本 V5.2

例9. 卡通插画

绘画主体

A girl holding a bunny, whimsical cartoon characters, many flowers around
一个抱着兔子的小女孩，异想天开的卡通人物，周围有很多鲜花

场景

White background, illustration, warm color palette
白色背景，插画，暖色调

风格

Henri Charles Manguin's style, Indonesian art
亨利·夏尔·芒更风格，印尼艺术

画质

HD
高清

基础设置

--ar 2:3 --niji 5
画面比例 2:3　版本 Niji5

例 10. 像素人物插画

绘画主体
Cartoon characters stand in a line
卡通人物们站成一排

场景
Kawaii, cute and dreamy, clean background
可爱的，可爱而梦幻，干净的背景

风格
Pixel art
像素艺术

基础设置
--ar 3:2 --niji 5
画面比例 3:2　版本 Niji5

如果生成图像的像素效果不强，可以尝试在风格关键词后增加权重多试几次。

例 11. 游戏场景设计插画

绘画主体

A small island with a wooden house and palm trees, beach scenes
一个有木屋和棕榈树的小岛,海滩边

场景

Xbox live preview, photo realistic technique, terracotta, textures
游戏 Xbox live 预览图,逼真的技术,陶土风,有纹理的

风格

Nintendo style, vintage aesthetics
任天堂风格,古典美学

画质

8K quality
8K 画质

基础设置

--ar 13:6 --v 5.2
画面比例 13:6 版本 V5.2

例 12. 赛博朋克插画

绘画主体
Deserted city buildings, future city, cyberpunk characters, many towering skyscrapers, neon cold lighting, unreal engine, futuristic shops and bars, crowd, hovering billboards, glass architecture, high-tech devices
废弃的城市建筑，未来城市，赛博朋克风的角色，很多摩天大楼，霓虹灯冷光，虚幻引擎，未来主义的商店和酒吧，人群，盘旋的广告牌，玻璃建筑，高科技的设备

场景
Ultra wide angle city, ultra wide shot, raining day, sharp focus, sharp light, vivid colors
超广角城市，超广角镜头，下雨天，锐利的焦点，锐利的光线，生动的颜色

风格
Cyberpunk, photorealistic
赛博朋克，逼真

画质	基础设置
16K quality	--ar 3:4 --v 5.2
16K 画质	画面比例 3:4
	版本 V5.2

例 13. 涂鸦插画

绘画主体

A cartoon girl wearing a T-shirt and sneakers, long hair, kittens on the ground, coloured background

一个穿着 T 恤和运动鞋的卡通女孩，长长的头发，地上有小猫，彩色的背景

场景

Clear illustration, bold lines and solid colors, mixed patterns, emoji installations

清晰的插图，粗线条和纯色，混合图案，表情符号的装置艺术

风格

Graffiti, in the style of Keith Haring, in the style of grunge beauty

涂鸦，凯斯·哈林风格，颓废美学

基础设置

--niji 5

版本 Niji5

例 14. 彩色铅笔插画

绘画主体

Portrait close-up of a young European girl, elegant and charming temperament

一位欧洲少女的肖像特写，优雅而迷人的气质

风格

Colored pencil drawing, sketch works with realistic super detail portrait style, super realistic sketch, line art

彩色铅笔画，具有逼真细节的肖像风格素描作品，超真实的素描，线条艺术

画质	基础设置
HD	--v 5.2
高清	版本 V5.2

例 15. 漫画分镜插画

绘画主体
A girl with black hair
黑头发的女孩

场景
Separate scene, vivid lighting and shadows, sakura in the background
分别的场景，生动的灯光和阴影，背景有樱花

风格
Japanese manga page, different groups of storyboards, magazine, girl manga
日漫风格页面，几组不同的分镜，杂志，少女漫画

画质
8K quality, OC rendering
8K 画质，OC 渲染

基础设置
--ar 9:16 --niji 5
画面比例 9:16 版本 Niji5

例 16. 水彩插画

绘画主体
A flower shop on the old city street, street scenes
一个古老的城市街道上的花店,街景

场景
Soft and dreamy atmosphere, light-filled outdoor scenes, side view, bright, dreamy
柔和而梦幻的氛围,充满光的室外场景,侧视图,明亮的,梦幻的

风格
Watercolor paint, in style of Yuko Nagayama
水彩画,永山裕子风格

基础设置
--niji 5
版本 Niji5

例 17. 乡村风景插画

绘画主体
Countryside, clouds, old houses, green fields, mountains
乡村，云朵，复古建筑，绿色的田野，山

场景
Natural light, asymmetrical composition
自然光，不对称构图

风格
In the style of Makoto Shinkai, Dima Dmitriev, anime art, cute and dreamy
新海诚风格，迪玛·德米特里耶夫风格，动漫艺术，可爱而梦幻

画质	基础设置
HD	--niji 5
高清	版本 Niji5

例 18. 扁平风格插画

绘画主体
Snowy mountain
雪山

场景
Wide angle, scenery, natural light, sky in the background
广角，风景，自然光，背景是天空

风格
Minimalism, in the style of figurative minimalism, organic shapes and lines, danish design, flat illustration
极简主义，具象极简风格，有机的形状和线条，丹麦设计，平面插图

画质
8K quality
8K 画质

基础设置
--ar 3:4 --niji 5
画面比例 3:4　版本 Niji5

例 19. 穆夏风格插画

绘画主体

European style maiden, single figure, more details on the face, massive floral decoration, star and moon glitter, gold dust, rose, lots of gold leaves, Byzantine decoration, delicate face, extreme details

欧式少女，单一人物，更多的脸部细节，大量的花卉装饰，闪闪发光的星星和月亮，金粉，玫瑰，许多金色的树叶，拜占庭装饰，精致的脸，极致的细节

场景

European court, garden, golden light, gold shimmering details, asymmetrical composition, very bold outline, clean contour lines, bright watercolor effect

欧洲宫廷，花园，金色的光，金光闪闪的细节，不对称构图，非常醒目的轮廓，干净的轮廓线，明亮的水彩

风格

In the style of Alphonse Mucha, Baroque, Art Nouveau, Tarot cards, Rococo art style, Impressionism

穆夏风格，巴洛克式，新艺术风格，塔罗牌，洛可可艺术风格，印象派

画质	基础设置
8K quality 8K 画质	--niji 5 版本 Niji5

例 20. 宫崎骏风格插画

绘画主体

A kid on a bike, balloons on the bike, look bright and cheerful, body movements playful, smiling face

骑车的小孩,带气球的自行车,神情灵动开朗,肢体动作俏皮,微笑的脸庞

场景

Field, morning light, the background is blue sky and white clouds, asymmetrical composition

田野,晨光,背景是蓝天白云,不对称构图

风格	**画质**	**基础设置**
Miyazaki Hayao style, studio Ghibli, dream style	HD	--niji 5
宫崎骏风格,吉卜力工作室风格,梦幻风格	高清	版本 Niji5

例 21. 哥特式风格插画

绘画主体

Gorgeous dark maiden, lace skirt, exquisite facial features, ultra-detailed
华丽的黑暗少女，蕾丝裙，精致的五官，超细节

场景

Retro dark vintage, church background, chiaroscuro, low angle perspective, cinematic bottom light, bokeh, soft focus, medium shot
暗黑风，教堂背景，明暗对比，低角度透视，电影底光，散景，软焦，中景

风格	画质	基础设置
Gothic gloomy	HD	--v 5.2
哥特式黑暗	高清	版本 V5.2

例 22. 街景图片插画

绘画主体

The city in the rain, crowd, neon-lit urban style, city street, cityscape, misty blue, tall buildings by the street

雨中的城市，人群，霓虹灯的城市风格，城市街道，城市景观，雾蒙蒙的蓝色，街旁有高楼

风格

Modern impressionism, in the style of Evgeny Lushpin

现代印象派，叶夫根尼·卢什平风格

画质	基础设置
HD	--niji 5
高清	版本 Niji5

例 23. 日式漫画风格插画

绘画主体
A princess-like girl is walking on an ancient bridge over the lake, lantern
一位公主般的少女走在湖中的古桥上，灯笼

场景
In the moonlit night
月色皎洁的夜晚

风格
Style of Reiko Shimizu
清水玲子风格

画质
HD
高清

基础设置
--niji 5
版本 Niji5

例 24. 卡通风格插画

绘画主体
A cartoon style bus
一辆卡通样式的公共汽车

场景
White background, illustration
白色背景，插画

风格
Style of Taro Gomi
五味太郎风格

画质	基础设置
HD 高清	--niji 5 版本 Niji5

例 25. 抽象线条风格插画

绘画主体
Moving lines make up mountains
动线组成的山脉

风格
Moving lines, ripples, hypnosis, blueprint, Abstractionism
动线，涟漪，催眠，蓝图，抽象艺术

画质	基础设置
8K quality 8K 画质	--niji 5 版本 Niji5

例 26. 简笔插画

绘画主体
A cute little boy with a dog, full body
一个可爱的小男孩带着一只小狗，全身

风格
Minimalist, style of Keith Haring, hand-drawn illustration
极简主义，凯斯·哈林风格，手绘插画

场景
Clean background, bold lines and solid colors, style expressive, line drawing
干净的背景，大胆的线条和纯色，风格表现力强，线条绘画

画质
HD
高清

基础设置
--ar 3:4 --niji 5
画面比例 3:4　版本 Niji5

例 27. 丙烯装饰插画

绘画主体
Colorful painting of plants
五颜六色的植物

场景
Abstract, dripping, impasto, thick brush strokes, three-dimensional quality, bright color
抽象，滴落，浓墨重彩，笔触厚重，三维质量，明亮的颜色

风格
Acrylic painting, in the style of palette knife, hard edge painting, impressionism
丙烯画，调色刀风格，硬边画，印象派

画质	基础设置
8K quality	--ar 3:4 --v 5.2
8K 画质	画面比例 3:4　版本 V5.2

例 28. 童话绘本插画

绘画主体
Wooden house in the forest, a huge tree, fantastical creatures, magical landscapes, intricate details, enchanted forest
森林中的木屋，一棵巨大的树，梦幻般的生物，神奇的风景，复杂的细节，魔法森林

场景
Varied colors, dreamy atmosphere
多种色彩，梦幻的氛围

画质
16K quality
16K 画质

基础设置
--ar 16:9 --v 5.2
画面比例 16:9　版本 V5.2

风格
Fairy tale illustration style, hand-drawn style
童话故事插图风格，手绘风格

Tips

Q：在绘本场景出现人物的情况下，如何保持画面中人物形象的统一？

A：panels 命令可以生成不同的或连续性的动作。只要在命令后面加入"X panels with different poses+ 动词"或"X panels with continuous+ 动词"（X 表示需要的数值），即可得到连续性的人物图像，帮助进行绘本的创作。

例 29. 点彩插画

绘画主体
Natural scenery, forest, plant, trees and flowers
自然风景，森林，植物，树木和花朵

画质
8K quality
8K 画质

场景
Soft and dreamy feel, bright and pastel colors, slightly blurred edges
柔和而梦幻的感觉，色调明亮且温柔，边缘模糊

基础设置
--ar 3:4 --v 5.2
画面比例 3:4　版本 V5.2

风格
Pointillism, colorful dots, impressionist style, the style of Monet, oil painting texture
点彩风格，彩色圆点，印象派风格，莫奈风格，油画质感

例 30. 写实主义插画

绘画主体

A sparkling stream, rocks on the shore, sunsets, mists, autumn leaves, landscape photo
波光粼粼的溪流，岸边有石块，落日，迷雾，秋天的落叶，风景照片

场景

Sharp focus, highly detailed, insanely detailed, high quality, high resolution
锐利的焦点，细节丰富，疯狂的细节，高品质，高分辨率

风格

Ultra-realistic, photo realistic, photo realism, photorealistic, intricate details
超写实，照片级写实，照片写实主义，逼真，复杂的细节

画质	基础设置
8K quality	--ar 2:3 --v 5.2
8K 画质	画面比例 2:3　版本 V5.2

例1. 水果胶带设计

绘画主体
Masking paper tape with many fruit elements, fine details
美纹纸水果元素胶带，精细细节

风格
Fairy book illustration style, clean background, fresh and natural style
童话书插画风格，干净的背景，清新自然风格

场景
Close-up, fantastic, Kawaii, global lighting, product view
特写，梦幻，卡哇伊（可爱），全局照明，产品视图

画质
4K quality
4K 画质

基础设置
--v 5.2
版本 V5.2

例 2. 植物胶带设计

绘画主体

Flowers washi tapes, in the style of intricate, delicate flower and garden paintings
花卉和纸胶带（编者注，和纸是纸的一种类型），采用复杂、精致的花卉和花园绘画风格

场景

Atmosphere of dreamlike quality
梦幻般的氛围

风格	画质	基础设置
Barbizon school, tondo, cloisonnism, raw materials 巴比松画派，圆形的绘画或浮雕，景泰蓝，原材料	HD 高清	--v 5.2 版本 V5.2

例 3. 植物贴纸设计

绘画主体

A set of stickers, featuring watercolor illustrations of fruits and flowers on a table, in the style of red and bronze
一组贴纸放在桌子上,贴纸上有水果和鲜花的水彩插图,红色和青铜色

场景

Romanticized country life, books and portfolios
浪漫的乡村生活,书籍和作品集

风格

white and light red, assemblage of maps
白色和浅红色,地图组合

画质	基础设置
HD	--ar 3:4 --v 5.2
高清	画面比例 3:4 版本 V5.2

例4. 书签设计

绘画主体

A metal tag with flowers and birds
带有花与鸟的金属标签

场景

Global illumination，grand scale
全局照明，规模宏大

风格

Chinese New Year celebration style, light gold and sky blue, sea blue, paper cutting, handmade design
中国新年庆祝风格，浅金色和天蓝色，海蓝色，剪纸，手工设计

画质	基础设置
4K quality	-- v 5.2
4K 画质	版本 V5.2

例 5. 捕梦网设计

绘画主体
Dreamcatcher, hand-woven, natural feathers, natural green rough stone, LED lighting, exquisite alloy accessories
捕梦网，手工编织，天然羽毛，天然绿色原石，LED 灯光，精美合金配件

场景
Black background, global illumination, panorama, symmetrical composition, e-commerce product view
黑色背景，全局照明，全景，对称构图，电商产品视图

风格
Realistic photography style, standard lens
写实摄影风格，标准镜头

画质	基础设置
Ultra HD，32K quality 超高清，32K 画质	--v 5.2 版本 V5.2

例 6. 立体贺卡设计

绘画主体
Three-dimensional greeting cards
立体贺卡

风格
Handmade, children's crafts
手工制作，儿童手工

场景
Open flowers and blue ferris wheel, symmetrical composition, close-up, bright
开放的鲜花和蓝色摩天轮，对称构图，特写，明亮

画质
Standard lens, depth of field effect, 32K quality
标准镜头，景深效果，32K 画质

基础设置
--v 5.2
版本 V5.2

例 7. 蝴蝶胸针设计

绘画主体
Butterfly brooch
蝴蝶胸针

场景
Extensive use of palette knives, Rococo pastels, gauzy fabrics, light silvers and aquas, textured paint planes, product view
大量使用调色刀，洛可可粉彩，薄纱织物，浅银色和浅绿色，纹理油漆平面，产品视图

风格
Romantic style, fresh and elegant
浪漫主义风格，清新优雅的

画质	基础设置
32K UHD	--v 5.2
32K 超高清	版本 V5.2

例 8. 花盆设计

绘画主体
Flower pot, ceramic, hollow, creative flower pot, lotus leaf shape
花盆，陶瓷，镂空，创意花盆，荷叶造形

场景
Front view, natural light, fine details, clean background
正视图，自然光，精细细节，干净的背景

风格
Simplicity, master handwork, craftsman spirit
简约，精湛手工艺，工匠精神

画质
Realistic photography, 32K quality
写实摄影，32K 画质

基础设置
--v 5.2
版本 V5.2

例 9. 奶油胶手机壳设计

绘画主体

A mobile phone case made of cream glue, decorated with ice cream, rainbows, clouds and other elements
奶油胶制作的手机壳，用冰淇淋、彩虹、云彩等元素装饰

场景

Rich colors, bright light, product view, light background
色彩丰富，光线明亮，产品视图，浅色背景

风格

Cute style, full of childishness, harmonious colors
风格可爱，充满童趣，和谐的配色

画质	基础设置
Real photography, 8K quality 真实摄影，8K 画质	--v 5.2 版本 V5.2

例 10. 掐丝珐琅台灯设计

绘画主体
Enamel, table lamp, gorgeous, blue, green, red, warm light
搪瓷，台灯，华丽的，蓝、绿、红、暖光

场景
Mosaic elements, asymmetrical shapes
马赛克元素，不对称造型

风格
Rococo style, retro style
洛可可风格，复古风格

画质
Realistic photography, 8K quality
写实摄影，8K 画质

基础设置
--v 5.2
版本 V5.2

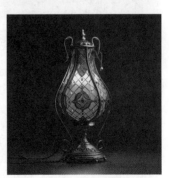

例 11. 水晶风铃设计

绘画主体
Wind chimes, small glass crystals, suspended on a round metal frame, colorful and harmoniously, arranged in, light gold, light cyan and magenta
风铃，小玻璃水晶，悬浮在圆形金属框架上，色彩缤纷，排列和谐，浅金色、浅青色和洋红色

场景
The warm sunlight shines on it, exquisite details, ambient light, golden ratio, sharp focus, soft light
和煦的阳光照射在上面，精致的细节，环境光，黄金画面比例，锐利的焦点，柔和的光线

风格	画质	基础设置
Modern design style	Real photography, 8K quality	--v 5.2
现代设计风格	真实摄影，8K 画质	版本 V5.2

例 12. 糖画设计

绘画主体
Translucent caramel lollipop, phoenix shape, flat
半透明的焦糖棒棒糖，凤凰形状，扁平

场景
Sugar dissolving process, traditional Chinese sugar painting, caramel color, translucent, thin, tough
溶糖工艺，中国传统糖画，焦糖色，半透明，薄，韧

风格
Traditional Chinese sugar painting style, handmade by masters, handicrafts
中国传统糖画风格，大师手作，手工艺品

画质
Realistic photography, 4K quality
写实摄影，4K 画质

基础设置
--v 5.2
版本 V5.2

例 13. 糖霜饼干设计

绘画主体
Frosted biscuits with different patterns, with frosted texture, depict French pastoral scenery, green, blue, brown
不同图案的糖霜饼干，有磨砂质感，描绘了法式田园风光，绿色，蓝色，棕色

场景
Summer atmosphere, simple brushstrokes, soft lighting, bright colors, simple lines, the overall visual beauty is soft, dreamy quality, delicate and soft tones and gentle brushstrokes
夏日气息，简洁的笔触，柔和的灯光，明亮的色彩，简洁的线条，整体视觉美感柔和，梦幻的品质，细腻柔和的色调和温柔的笔触

风格
Romantic pastoral style, baking art
浪漫的田园风格，烘焙艺术

画质
Realistic photography, close-up, 32K quality
写实摄影，特写，32K 画质

基础设置
--v 5.2
版本 V5.2

例 14. 藤编猫窝设计

绘画主体	**场景**
Pet cradle, rattan 宠物摇篮，藤条	Home use, placed in the living room, morning light, clean background, front view 家用，放在客厅，晨光，干净的背景，正视图

风格	**画质**	**基础设置**
Realistic photography 写实摄影	Standard lens, 32K quality , HD 标准镜头，32K 画质，高清	--ar 5:3--v 5.2 画面比例 5:3　版本 V5.2

例 15. 衍纸艺术设计

绘画主体

Paper quilling art, Van Gogh's Starry Night, wooden frame, clear layers, smooth lines, white space

衍纸艺术，梵高的《星夜》，木框，层次分明，线条流畅，留白空间

场景

Soft polish, light background, rich details

柔和抛光，浅色背景，细节丰富

风格	画质	基础设置
Master's handwork 大师手作	Product view, realistic photography 产品视图，写实摄影	--ar 5:3 --v 5.2 画面比例 5:3　版本 V5.2

第 3 章 DIY 设计

例 16. 纸杯定制设计

绘画主体
Disposable paper cup, with cartoon pattern on the side, castle in the sky, flying birds
一次性纸杯,侧面有卡通图案,天空之城,飞翔的小鸟

场景
Clean background, global lighting, product view
干净的背景,全局照明,产品视图

风格
Studio Ghibli, cute
吉卜力工作室风格,可爱

画质
32K Ultra HD
32K 超高清

基础设置
--v 5.2
版本 V5.2

例 17. 月饼礼盒设计

绘画主体

Gift paper box, brand design, art decoration, Chinese Song Dynasty landscape painting, light beige and light blue color scheme
礼品纸盒，品牌设计，艺术装饰，中国宋代山水画，浅米色和浅蓝色配色方案

场景

Global illumination, light background, front view
全局照明，浅色背景，正视图

风格

Graphic design, paper, nature, hyper-real
平面设计，纸张，自然，超真实

画质

3D rendering, OC renderer, 8K quality, HD
三维渲染，OC 渲染器，8K 画质，高清

基础设置

--ar 5:3 --v 5.2
画面比例 5:3　版本 V5.2

第 3 章　DIY 设计

例 18. 折纸犀牛设计

绘画主体
Origami, rhinoceros, gray paper on wooden table, with origami tools next to it
折纸，犀牛，灰色纸放在木桌上，旁边有折纸工具

场景
Morning light, close-up, asymmetrical composition
晨光，特写，不对称构图

风格
Realistic photography style
写实摄影风格

画质
Medium format camera, standard lens
中画幅相机，标准镜头

基础设置
--ar 4:3 --v 5.2
画面比例 4:3　版本 V5.2

例 19. 植物火漆印章设计

绘画主体
Bay leaf pattern wax seal, printed on envelope, green and gold foil
月桂叶图案蜡封，印在信封上，绿色和金色的箔纸

场景
Soft light, autumn atmosphere, golden ratio, exquisite details
光线柔和，秋天的气氛，黄金画面比例，精致的细节

风格
Noble and elegant style, British flavor, romanticism
高贵优雅的风格，英伦气息，浪漫主义

画质
realistic photography, 32K quality
写实摄影，32K 画质

基础设置
--ar 4:5 --v 5.2
画面比例 4:5　版本 V5.2

例 20. 珠绣灯设计

绘画主体
Beaded lantern, woven from handmade glass beads and blue beads
串珠灯笼，由手工玻璃珠和蓝色珠子编织而成

场景
Adopting complex weaving style, traditional Chinese palace lanterns, incredible details, exquisite details, bright light, super details, complex patterns, high-end texture, noble and elegant temperament, front view
采用繁复的编织风格，中国传统宫灯，令人难以置信的细节，精致的细节，明亮的光线，超级细节，复杂的图案，高端质感，高贵优雅的气质，正视图

风格
Ancient Chinese handicraft style
中国古代手工艺风格

画质
Realistic photography, 32K quality
写实摄影，32K 画质

基础设置
--v 5.2
版本 V5.2

例 21. 桌面风扇设计

绘画主体

Green desktop fan with yellow blades on a wooden table, with lemons on the table
木桌上有黄色叶片的绿色台式风扇,桌上有柠檬

场景

Global soft lighting, product perspective, summer atmosphere, central composition
全局柔和的灯光,产品视角,夏季氛围,中心构图

风格

Simple design style
简单的设计风格

画质

Photorealistic photography, detailed details, 4K quality
照片写实摄影,详细的细节,4K 画质

基础设置

--v 5.2
版本 V5.2

例 22. 钩针童鞋设计

绘画主体
Yellow, beige, blue crocheted baby slippers
黄色、米色、蓝色钩针婴儿拖鞋

风格
Animal prints, mainly installations
动物印花，主要是装置艺术

场景
The styles are dark blue, red, dark white, and dark pink, using precious materials, light gold, light brown, clean background, global illumination
款式有深蓝、红、深白、深粉，采用名贵材料，浅金、浅棕，干净的背景，全局照明

画质
HD
高清

基础设置
--v 5.2
版本 V5.2

例 23. 玉雕设计

绘画主体
Natural green landscape jade carvings, Chinese jade carvings, peaks, trees, pavilions, Chinese architecture
天然绿色山水玉雕，中国玉雕，山峰，树木，亭台楼阁，中式建筑

场景
100mm f/2.8 2x Super Macro, global Illumination, product view
100mm f/2.8 2x 超微距，全局照明，产品视图

风格
Old pit style, made of crystal emerald
老坑风格，水晶翡翠制成

画质
HD
高清

基础设置
--v 5.2
版本 V5.2

例 24. 剪纸艺术设计

绘画主体
Paper-cut art, flat, paper-cut, pavilion, surrounded by peony flowers, two people drinking tea

剪纸艺术，平面，剪纸，凉亭，牡丹花簇拥，两个人喝茶

场景
Clean background, global illumination, product view

干净的背景，全局照明，产品视图

风格
Graphic design, red

平面设计，红色

画质	基础设置
4K quality	--v 6.0
4K 画质	版本 V6.0

例 25. 定制帆布包设计

绘画主体

New cotton shopping bag, watercolor painting, traditional Chinese landscape style, strange rocks, bamboo, orchids

新款棉布购物袋，水彩画，中国传统山水风格，奇石，竹子，兰花

场景

Clean background, global lighting, product view, soft light, detailed details

干净的背景，全局照明，产品视图，柔和的光线，详细的细节

风格

Pleasing
令人愉悦的

画质	基础设置
32K UHD images	--v 5.2
32K 超高清图像	版本 V5.2

例 26. 盆栽摆件设计

绘画主体
Two bonsai trees
两棵盆景树

场景
Made of crystals, spiral group, dark gold and gray, dreamlike installations, gravity-defying landscapes, made of liquid metal, black background
由水晶制成，螺旋群，暗金色和灰色，梦幻般的装置，反重力景观，液态金属制成，黑色背景

风格
In the style of fantastical contraptions
奇幻装置的风格

画质	基础设置
32K UHD images	--v 5.2
32K 超高清图像	版本 V5.2

例 27. 城市拼图设计

绘画主体
Jigsaw puzzles, small town scenery, busy residents
拼图游戏，小镇风光，忙碌的居民

场景
New mosaics, interesting life scenes, intricate storytelling, soft lighting, product display
新颖的马赛克，有趣的生活场景，错综复杂的联系，柔和的灯光，产品展示

风格
Maximalist style, graffiti art
极繁主义风格，涂鸦艺术

画质	基础设置
32K UHD images	--ar 16:9 --v 5.2
32K 超高清图像	画面比例16:9　版本V5.2

例 28. 充电宝设计

绘画主体
Power bank design, pink and white, intricate filigree, cartoon sticker
移动电源设计，粉色和白色，复杂的花丝，卡通贴纸

场景
Incredible details, front view, clean background
令人难以置信的细节，正视图，干净的背景

风格
Realistic photography style, product view
写实摄影风格，产品视图

画质	基础设置
4K Ultra HD	--v 5.2
4K 超高清	版本 V5.2

例 29. 刺绣图设计

绘画主体

Embroidery, blooming peonies and playful butterflies, reds, pinks and greens, lots of white space, detailed embroidery textures
刺绣，盛开的牡丹和嬉戏的蝴蝶，红色，粉色和绿色，大量的空白，细致的刺绣纹理

场景

Clean background, golden section, good proportions, panorama, front view
干净的背景，黄金分割，良好的画面比例，全景，正视图

风格

Elegant and noble Chinese style, ultra-detailed
优雅高贵的中国风，超细节

画质

Photoreal, ultra-detailed, realistic photography, 32K quality
照片真实，超细节，写实摄影，32K 画质

基础设置

--v 5.2
版本 V5.2

例 30. 蛋糕设计

绘画主体
Cake, decorated like a green forest, has villages and houses in it
蛋糕，装饰得像一个绿色森林，里面有村庄和房子

场景
Simple, white tabletop cake, morning light
简约，白色桌面蛋糕，晨光

风格
Lo-fi aesthetic style
低保真美学风格

画质
8K standard lens, full frame camera, central composition, realistic photography, product display
8K 标准镜头，全画幅相机，中心构图，写实摄影，产品展示

基础设置
--v 5.2
版本 V5.2

例 31. 滴胶戒指设计

绘画主体
Epoxy ring with floral landscape inside, intricate, delicate flowers
内有花卉景观的环氧环,复杂,精致的花朵

场景
Delicate rendering, serene garden landscape, charming color landscape, white background, front view, golden section, symmetry
细腻的渲染,宁静的园林景观,迷人的色彩景观,白色背景,正视图,黄金分割,匀称

风格
Simple design, Monet style, romanticism
简约设计,莫奈风格,浪漫主义

画质
Realistic photography, 32K quality
写实摄影,32K 画质

基础设置
-- ar 8:10 --v 5.2
画面比例 8:10　版本 V5.2

第 4 章
实办设计

例1. 手办基础设计

绘画主体
A Model Kit, an anime girl with base
一个手办,一个有底座的动漫女孩

场景
Full length shot, white background
全身视角,白色背景

风格
In the style of red and light maroon, cute and colorful, charming anime characters
红色和浅栗色的风格,可爱多彩的,充满吸引力的动漫角色

画质
Surreal photography
超现实摄影

基础设置
--v 5.2
版本 V5.2

例 2. 积木手办设计

绘画主体

A model kit of a cute architecture, LEGO bricks
一个可爱的建筑手办，乐高积木

场景

Isometric view, multidimensional shading
等距视图，多维阴影

风格

Playful cartoons, small and exquisite design
有趣的卡通，小巧精致的设计

画质

Hyper quality
超高质量

基础设置

--v 5.2
版本 V5.2

例 3. 宇航员手办设计

绘画主体
A cute model of astronaut sitting on a planet, PVC material
坐在星球上的可爱宇航员模型，PVC 材质

场景
In the universe
在宇宙中

风格
Cute and dreamy, clean and simple designs, robotics kids, celestialpunk
可爱梦幻，干净简洁的设计，机器人小孩风格，天体朋克风格

画质
Ultra-realistic
超写实

基础设置
--v 5.2
版本 V5.2

例 4. 手办三视图设计

绘画主体
A little boy in a tracksuit and a backpack, cheerful and lovely
一个穿着运动服背着背包的小男孩，开朗可爱

场景
No dividing line, full length shot, generate three views, namely the front view, the back view and the side view
无分割线，全身镜头，生成三视图，即正视图、后视图和侧视图

风格
Pixar style, ultra details
皮克斯风格，超级细节

画质
Octane rendering
Octane 渲染

基础设置
--ar 16:9 --niji 5
画面比例 16:9　版本 Niji5

例 5. 迷你厨房手办设计

绘画主体
Miniature food kitchen，doll house toy, mini cooking kitchen
微型食品厨房，娃娃屋玩具，迷你烹饪厨房

场景
Organized chaos，medium long shot
有组织的混乱，中远景

风格
Realistic and hyper-detailed renderings, photorealistic paintings, stark contrast, contemporary DIY
写实与超细节渲染，写实绘画，鲜明对比，当代DIY

画质
10K quality
10K 画质

基础设置
--v 5.2
版本 V5.2

第4章 手办设计

例6. 机械蜻蜓手办设计

绘画主体

A dragonfly made from a metal mechanical mechanism, leg standing, kinetic installations
一只由金属机械结构制成的蜻蜓，腿站立，动能装置

场景

Panoramic view, aerial photography
全景，航空摄影

风格

Machine aesthetics, dieselpunk, illusionistic details, mechanical realism
机器美学，柴油朋克，幻觉细节，机械写实主义

基础设置

--v 5.2

版本 V5.2

例 7. 恐龙手办设计

绘画主体

A dinosaur model toy, standing on a rock, PVC material, dark green and camouflage, precise limbs
恐龙模型玩具，站在岩石上，PVC 材质，墨绿迷彩，四肢精准

场景

Long shot, full length shot, place on a clean table top
长镜头，全身镜头，放置在干净的桌面上

风格

In the style of collecting and modes of display, mori kei, accurate, detailed
收藏风格和展示方式，森系，准确，细致

画质

8K resolution
8K 分辨率

基础设置

--ar 16:9 --v 5.2
画面比例 16:9　版本 V5.2

例 8. 金属手办设计

绘画主体

A warrior with outstretched wings, metal material, with a circular base
张开翅膀的武士，金属材质，圆形底座

场景

Volumetric lighting, wide and exquisite, white background
体积光，宽阔精致，白色背景

风格
Super realistic
超写实

画质
3D render
三维渲染

基础设置
--v 5.2
版本 V5.2

例 9. 生肖鼠手办设计

绘画主体

Miniature sculptures of a cute little mouse, holding an apple, clay material

可爱小老鼠的微型雕塑，拿着苹果，黏土材质

场景

Front view, high contrast background

正视图，高对比度背景

风格

Soft colors, child-like innocence, near physical, Pixar style

柔和的色彩，童心，接近实物，皮克斯风格

画质	基础设置
32K UHD	--v 5.2
32K 超高清	版本 V5.2

例 10. 帆船手办设计

绘画主体

A wooden boat, wool felt material, light teal and orange, miniature sculptures, use of fabric, cheerful colors
木船，羊毛毡材质，浅青色和橙色，微型雕塑，布料的运用，欢快的色彩

场景

Natural lighting, close-up view, blurred background
自然采光，特写镜头，背景模糊

风格	画质	基础设置
Simple and cute	Shot on 70mm	--v 5.2
简单又可爱	70mm 镜头拍摄	版本 V5.2

例 11. 编织考拉设计

绘画主体

An adorable koala, knitted and crocheted, sitting in a tree
一只可爱的考拉，编织和钩针，坐在树上

场景

Front lighting, Australian landscapes
正面照明，澳大利亚风景

画质

Soft-focus technique
软聚焦技术

基础设置

--v 5.2
版本 V5.2

例12. 木质建筑手办设计

绘画主体
Model of a wooden house with garden, wood material, with a wooden base, the original wood color
带花园的木屋模型，木质材质，带木质底座，原木色

场景
Clean background, Isometric view
干净的背景，等距视图

风格
Romanticized country style, traditional Chinese architectural style, delicate constructions
浪漫乡村风格，中国传统建筑风格，精致的构造

基础设置
--v 5.2
版本 V5.2

画质
High resolution
高分辨率

例 13. 汉堡手办设计

绘画主体
A mini burger, wool felt material, fluffy, close to the real color, dinky and delicate
一个迷你汉堡，羊毛毡材质，蓬松，接近真实颜色，小巧精致

场景
Place on a plate, soft illumination
放在盘子上，柔和的照明

画质
In focus, 8K quality
对焦清晰，8K 画质

基础设置
--v 5.2
版本 V5.2

例 14. 存钱罐设计

绘画主体
Piggy shaped piggy bank, ceramic material, light pink and light gold, strong facial expression, round and cute, with an opening on the back
小猪形状的存钱罐，陶瓷材质，浅粉色和浅金色，强烈的面部表情，圆润可爱，背面有开口

场景
Isometric view, placed on a clean table
等距视图，放置在干净的桌子上

风格
Ceramic texture, photo realistic rendering
陶瓷质感，真实感渲染

画质
3D rendering
三维渲染

基础设置
--v 5.2
版本 V5.2

例15. 汉服手办设计

绘画主体
Little girl in Hanfu, cute and elegance, traditional Hanfu, loose robes, exquisite ancient Chinese hair bun, elaborate and refined hairstyle
汉服小姑娘，可爱又优雅，传统的汉服，宽松的长袍，精致的中国古式发髻，精致脱俗的发型

场景
Full body, front view, bright background, depth of field
全身，正视图，明亮的背景，景深

风格
Pop Mart, blind box style, mind-blowing details
泡泡玛特，盲盒风格，细节震撼

画质	基础设置
UHD clarity	--v 5.2
超高清晰度	版本 V5.2

例 16. 桥梁模型手办设计

绘画主体

A model toy of bridge, tabletop decoration, structural symmetry
桥梁模型玩具，桌面装饰，结构对称

场景

Drone view, soft illumination
无人机视角，柔和的照明

风格

Precise lines and shapes, transavanguardia, realist details, eco-kinetic, lifelike renderings
精确的线条和形状，超前卫，逼真的细节，生态动力学，逼真的渲染

基础设置

--ar 16:9 --v 5.2
画面比例 16:9　版本 V5.2

例 17. 小天使手办设计

绘画主体

A little angel with wings, clay material, light purple hair and white wings, with bouquet of flowers

有翅膀的小天使,黏土材质,浅紫色的头发和白色的翅膀,拿着一束花

场景

Top light, product view
顶光,产品视图

风格

Studio Ghibli
吉卜力工作室风格

画质	基础设置
UHD high quality	--v 5.2
超高品质	版本 V5.2

例 18. 鲸鱼手办设计

绘画主体

Two kawaii baby whales, small in shape, tails up, blue and white, miniature sculptures, glossy

两只可爱的小鲸鱼，形状小巧，尾巴朝上，蓝色和白色，微型雕塑，有光泽

场景

Put them on a white soft cloth

将它们放在白色的软布上

风格	画质	基础设置
Animation stills, minimalist designs	HD	--v 5.2
动画剧照，极简设计	高清	版本 V5.2

例 19. 赛博小鸟手办设计

绘画主体
A bird with an angry expression, vivid facial expressions, standing
愤怒的小鸟，表情生动，站着

场景
In an industrial environment
在工业环境中

风格
Cyberpunk, life-like avian, illustrations
赛博朋克，栩栩如生的鸟类，插画

画质
Rendered in unreal engine, VRay
虚幻引擎渲染，VRay 渲染器

基础设置
--v 5.2
版本 V5.2

> 若想获取同一种风格的不同视角图像，可以尝试在关键词中添加 front view（正视图）、back view（后视图）和 side view（侧视图）。

例 20. 微缩模型设计

绘画主体

Mini figurines in strawberries, a lot of people are reclaiming the wilderness, miniature sculpture
草莓中的迷你雕像，很多人正在开垦荒地，微型雕塑

场景

Use of light background
使用浅色背景

风格

In a surreal fashion photography style, extreme close-up, minimalism style, soft color blend
超现实的时尚摄影风格，极致特写，极简风格，柔和的色彩混合

画质	基础设置
8K quality	--v 5.2
8K 画质	版本 V5.2

例 21. 气球小狗设计

绘画主体
A balloon puppy, dark white and light blue, balloon material, balloon doll
气球小狗，深白浅蓝，气球材质，气球公仔

场景
Outdoor scene
室外场景

风格
Multilayered dimensions, simplified dog figures, handmade, marvelous details
多层维度，简约狗造型，手工制作，精彩细节

画质	基础设置
HD 高清	--v 5.2 版本 V5.2

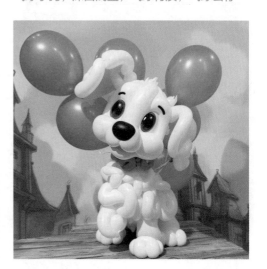

例 22. 木质猫手办设计

绘画主体
A carved wooden cat, in the style of adorable toy sculptures, lively action poses, light brown
木雕猫，可爱玩具雕塑风格，动作活泼，浅棕色

场景
Side view, delicate shading
侧视图，阴影细腻

风格
Minimalist, kawaii, raw and unpolished
极简，可爱，原始且未经雕琢

画质	基础设置
Highly textured 质感十足	--niji 5 版本 Niji5

例 23. 火箭手办设计

绘画主体
A LEGO rocket, with many parts on it
乐高火箭，上面有很多零件

场景
Clean background
干净的背景

风格
Realistic and hyper-detailed renderings, Dada-inspired constructions, detailed scientific subjects
逼真且超详细的渲染，达达风格的结构，详细的科学主题

画质
UHD image
超高清图像

基础设置
--v 5.2
版本 V5.2

例 24. 飞船手办设计

绘画主体
A model of Millennium Falcon, Star Wars collector's edition
千年隼模型，星球大战收藏版

风格
Boldly black and white, detailed ship sails, exquisite craftsmanship, three-dimensional puzzles
大胆的黑白，细致的船帆，精湛的工艺，立体的拼图

场景
Put on the table, global illumination
放在桌子上，全局照明

画质
Lifelike
栩栩如生

基础设置
--v 5.2
版本 V5.2

第 4 章 手办设计

例 25. 卫星手办设计

绘画主体
A satellite model, solar panels like wings, steel/iron frame construction
卫星模型，翅膀般的太阳能帆板，钢 / 铁框架结构

场景
Grandeur of scale, natural lighting
规模宏大，自然采光

风格
In the style of silver, white and gray, precision engineering
银白灰风格，白色和灰色，精密工程

画质
The best picture quality
最佳画质

基础设置
--v 5.2
版本 V5.2

例 26. 飞机手办设计

绘画主体
Civil aviation aircraft model, on a stand near windows
民航飞机模型，放在靠近窗户的展台上

场景
The background is blue sky and white clouds, natural lighting, super side angle
背景是蓝天白云，自然采光，超侧视角

风格
In the style of red and white, still shot and depth of field, collecting and modes of display
红白风格，静态镜头和景深，收集和展示方式

画质
Hyper-realistic, 8K quality
超逼真，8K 画质

基础设置
--v 5.2
版本 V5.2

第 4 章 手办设计

例 27. 直升机手办设计

绘画主体
A model of an army helicopter, with blue swooping wing
一架军用直升机模型，带有蓝色俯冲翼

场景
Bokeh
背景虚化

风格
High-tech feel, futuristic mechanical style, light silver red
高科技感，未来机械风格，浅银红色

画质
Official art, detailed, physical shooting
官方艺术，细节，实物拍摄

基础设置
--niji 5
版本 Niji5

123

例 28. 赛车手办设计

绘画主体
A racing car model, unreal engine, red, spray painted
赛车模型，虚幻引擎，红色，喷漆

风格
Mechanized abstraction, frostpunk, mecha anime, sharp car lines
机械化抽象，寒霜朋克，机甲动漫，锋利的汽车线条

画质
Cinema 4D rendering
电影 4D 渲染

基础设置
--v 5.2
版本 V5.2

例 29. 公交车手办设计

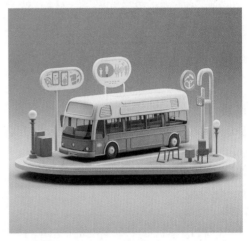

绘画主体
A model bus with stops and a sign, resin material
一个带有站牌的公交车模型，树脂材质

场景
Minimalist sets
极简布景

风格
Free-flowing lines, in the style of white and green, delicate and lovely, realistic details
线条流畅，白绿相间的风格，精致可爱，细节逼真

画质	基础设置
UHD image 超高清图像	--v 5.2 版本 V5.2

例 30. 摩托车手办设计

绘画主体
A model of motorcycle on the desktop
桌面摩托车模型

场景
Soft and bright light and shadow, close view, golden ratio
光影柔和明亮，近景，黄金画面比例

风格
Retroism, classic color scheme, classic American motorcycle, machine aesthetics, precise craftsmanship
复古主义，经典配色，经典美式摩托车，机器美学，做工精细

画质
High resolution
高分辨率

基础设置
--v 5.2
版本 V5.2

第 5 章
LOGO 设计

例 1. LOGO 基础设计

绘画主体
LOGO, house
标志，房子

场景
Simple, clean background, global illumination, central composition, front view
简单，干净的背景，全局照明，中心构图，正视图

风格
Graphic design
平面设计

画质	基础设置
4K quality	--niji 5
4K 画质	版本 Niji5

例2.LOGO个性化设计

绘画主体
Signs, books and mountains
标志，书籍和山

场景
Curves, black and white lines, clean background, global illumination, central composition, front view
曲线，黑白线条，干净的背景，全局照明，中心构图，正视图

风格
Minimalist art, graphic design
极简艺术，平面设计

画质	基础设置
HD	--niji 5
高清	版本 Niji5

例 3.LOGO 艺术化设计

绘画主体

LOGO, elephant and banana leaves, orange and green
标志，大象和香蕉叶，橙色和绿色

场景

Tropical atmosphere, clean background, global illumination, central composition, front view
热带氛围，干净的背景，全局照明，中心构图，正视图

风格

Picasso style, Dadaism, graphic design
毕加索风格，达达主义，平面设计

画质	基础设置
4K quality	--niji 5
4K 画质	版本 Niji5

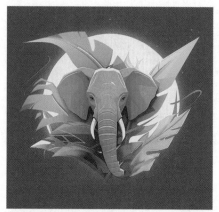

例 4. 参考图生成 LOGO 设计

垫图关键词：

绘画主体
Simple drawing, snowman
简笔画，雪人

场景
Children's painting, black and white lines
儿童画，黑白线条

风格
Minimalist style
极简风格

画质
HD
高清

基础设置
--niji 5
版本 Niji5

LOGO 关键词：

绘画主体
LOGO, colorful
标志，多彩

场景
Central composition, bright atmosphere
中心构图，明亮的气氛

风格
Graphic design, constructivism, typography style
平面设计，建构主义，印刷排版风格

画质
HD
高清

基础设置
--v 5.2
版本 V5.2

例 5. 字母 W 延展 LOGO 设计

绘画主体
LOGO design, butterfly and letter W, harmonious colors
标志设计，蝴蝶和字母 W，和谐的配色

场景
Clean background, Dadaism, metallic texture, dark background
干净的背景，达达主义，金属质感，深色背景

风格
Flat illustration
平面插画

画质	基础设置
4K quality	--v 5.2
4K 画质	版本 V5.2

例 6. 字母 Y 延展 LOGO 设计

绘画主体
Letter Y LOGO, green vine texture
字母 Y 标志，绿色藤蔓纹理

场景
Vector illustration, simple, clean background
矢量图，简单，干净的背景

风格
Flat design, plant texture
平面设计，植物纹理

画质	基础设置
4K quality	--v 5.2
4K 画质	版本 V5.2

例 7. 双鱼座 LOGO 设计

绘画主体

Two fish LOGO, Pisces LOGO design, pink and blue, head and tail intersecting
两条鱼标志，双鱼座标志设计，粉色和蓝色，头尾相交

场景

Central composition, clean background
中心构图，干净的背景

风格

Bauhaus style, cartoon, fantasy
包豪斯风格，卡通，梦幻

画质

4K quality
4K 画质

基础设置

--v 5.2
版本 V5.2

例 8. 巨蟹座 LOGO 设计

绘画主体
Crab LOGO, crab, pink and blue
螃蟹标志，螃蟹，粉色和蓝色

场景
Realistic details, cute, cartoon style, icon, vector, beautiful, super professional
逼真细节，可爱的，卡通风格，图标，矢量，漂亮，超专业

风格
Cartoon style
卡通风格

画质
4K quality
4K 画质

基础设置
--niji 5
版本 Niji5

例 9. 多边形艺术风 LOGO 设计

绘画主体
Deer head LOGO，black and brown
鹿头标志，黑色和棕色

场景
Elegant lines, soft, light background, icon, vector, beautiful, super professional
线条优雅，柔和，浅色背景，图标，矢量，精美，超专业

风格
Polygonal art, simple graphic design
多边形艺术，简约平面设计

画质	基础设置
HD	--v 5.2
高清	版本 V5.2

例 10. 冰淇淋 LOGO 设计

绘画主体

Ice cream vector LOGO with cherry elements, lemon yellow and pink colors
点缀樱桃元素的冰淇淋矢量标志，柠檬黄和粉色

场景

Visual harmony, vibrant atmosphere, smokey background
视觉和谐，充满活力的氛围，烟熏背景

风格

Vibrant graffiti art style, spray paint style
充满活力的涂鸦艺术风格，喷漆风格

画质

UHD images
超高清图案

基础设置

--v 5.2
版本 V5.2

例 11. 巴士 LOGO 设计

绘画主体
Bus LOGO, color cartoon, yellow and white
巴士车标，彩色漫画，黄白相间

场景
Sporty atmosphere, splash-ink style, strong dynamic, white background, vector graphics
运动气息，泼墨风格，动感强烈，白底，矢量图形

风格
Retro red, blue and yellow smooth comic style
复古的红蓝黄流畅漫画风格

画质	基础设置
Ultra HD pattern 超高清模式	--v 5.2 版本 V5.2

例 12. 比萨 LOGO 设计

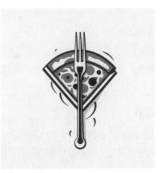

绘画主体
LOGO design, simple LOGO, pizza and fork
标志设计，简单标志，比萨和叉子

场景
Creative, super professional, award winning, light background, simple
创意，超专业，获奖，浅色背景，简单的

画质
HD
高清

风格
Illustration style, graphic design, hand drawn illustration
插画风格，平面设计，手绘插画

基础设置
--v 5.2
版本 V5.2

例 13. 法式复古风 LOGO 设计

绘画主体
Castle LOGO design surrounded by roses, intricate filigree entanglement
玫瑰包围的城堡标志设计，繁复的花丝缠绕

场景
Round stamp, light indigo blue atmosphere, Rococo style details, gorgeous vintage
圆形图章，浅靛蓝色氛围，洛可可风格细节，华丽复古

风格
Retro graphic design, graphic illustration
复古平面设计，平面插画

画质
4K quality
4K 画质

基础设置
--v 5.2
版本 V5.2

例 14. 花朵 LOGO 设计

绘画主体
Minimalist LOGO, simple LOGO, many tiled little houses filled with flowers
极简标志，简单的标志，许多瓷砖的小房子充满鲜花

场景
Creative, super professional, award-winning work, light background, lively colors, flat
创意，超专业，获奖作品，浅色背景，活泼的色彩，扁平化

风格	画质	基础设置
Flat design, mosaic style	4K quality	--niji 5
平面设计，马赛克风格	4K 画质	版本 Niji5

例 15. 徽标 LOGO 设计

绘画主体
Lions Athletic Club LOGO, dark orange and purple
狮子运动俱乐部标志，深橙色和紫色

场景
Sporty atmosphere, powerful atmosphere, dark background, central composition, vector illustration, colorful cartoon, realism and fantasy elements
运动氛围，力量感氛围，深色背景，中心构图，矢量插图，彩色漫画，写实主义与幻想元素

风格
LOGO sign style, graphic design
徽标标志风格，平面设计

画质
4K quality
4K 画质

基础设置
--niji 5
版本 Niji5

例 16. 男士头像 LOGO 设计

绘画主体
Man head LOGO holding metal tools, LOGO design, blue color, work clothes
手持金属工具的男子头像标志，标志设计，蓝色，工作服

场景
Central composition, clean background
中心构图，干净的背景

风格
American retro art, typography art
美式复古艺术，排版艺术

画质
4K quality
4K 画质

基础设置
--niji 5
版本 Niji5

例 17. 鸟头 LOGO 设计

绘画主体
Bird head LOGO, yellow and emerald
鸟头标志，黄色和翡翠色

场景
Sporty vibe, epic, award winning, cartoon core, vector illustration, white background, color harmony
运动氛围，史诗，获奖，卡通核心，矢量插图，白色背景，色彩和谐

风格
Flat design, flat style
平面设计，扁平化风格

画质	基础设置
8K quality	--niji 5
8K 画质	版本 Niji5

例18. 人体元素 LOGO 设计

绘画主体
Business LOGO design, human element, simple icon
商业标志设计，人体元素，简单的图标

场景
Simple brush strokes, suitable for vector art, including text design, no shadows or colors that are blurred or faded
简单笔触，适合矢量艺术，包括文本设计，无阴影或颜色模糊或褪色

风格
Minimalist art, graphic design
极简艺术，平面设计

画质	基础设置
4K quality	--v 5.2
4K 画质	版本 V5.2

例 19. 美式复古 LOGO 设计

绘画主体
Baseball boy business LOGO, LOGO design, simple icon, orange and blue
棒球男孩商业标志，标志设计，简单图标，橙色和蓝色

场景
Simple brush strokes, suitable for vector art, including text design, no shadows or colors blurred or faded, sporty atmosphere
简单笔触，适合矢量艺术，包括文本设计，无阴影或颜色模糊或褪色，运动氛围

风格
American retro style, American comics
美式复古风格，美式漫画

画质	基础设置
4K quality	--v 5.2
4K 画质	版本 V5.2

例 20. 剪影风格 LOGO 设计

绘画主体	场景
LOGO design, woman and olive leaf, mainly retro green and black	Simplicity, shaped canvas, organic materials, LOGO design, soft lines, projections
标志设计，女人和橄榄叶，复古绿色和黑色为主	简约，成型帆布，有机材料，标志设计，柔和的线条，投影

风格	画质	基础设置
Polygonal art style, minimalist flat design	4K quality	--v 5.2
多边形艺术风格，简约平面设计	4K 画质	版本 V5.2

例 21. 应用图标设计

绘画主体
App icon design, camera element, vibrant minimalist colors, retro camera
应用图标设计，相机元素，充满活力的简约色彩，复古相机

场景
Front view, vector illustration, 3D texture, light background
正视图，矢量图，三维纹理，浅色背景

风格
Graphic design, cute cartoon style
平面设计，可爱卡通风格

画质
4K quality
4K 画质

基础设置
--v 5.2
版本 V5.2

例 22. 元素图标设计

绘画主体
Fast food element icon, UI design, harmonious color combination
快餐元素图标，界面设计，和谐的色彩组合

场景
Tile image, illustration style, vector, illustration, light background
平铺图，插画风格，矢量，插画，浅色背景

风格
Flat design, cute style
平面设计，可爱风格

画质
4K quality
4K 画质

基础设置
--niji 5
版本 Niji5

第 6 章
海报主图设计

例 1. 立春海报主图设计

绘画主体
Birds are flying in the sky, green trees, mountain and water
小鸟飞在天空中，绿树，山水

场景
Early spring, green tone, blue sky
初春，绿色色调，蓝天

风格
Chinese ink painting, Chinese style
中国水墨画，中国风

画质
8K quality
8K 画质

基础设置
---ar 9:16 --v 5.2
画面比例 9:16　版本 V5.2

例 2. 芒种海报主图设计

绘画主体
A farmer in a straw hat, the farmer is working hard, smiling face
一位戴着草帽的农民，农民在辛勤劳作，微笑的脸

场景
The golden wheat field, autumn, blue sky, Chinese farmland, ultra-wide angle, highly detailed, high quality
金色的麦田，秋天，蔚蓝的天空，中国农田，超广角，高细节，高品质

风格
Chinese style, Chinese national trend illustration style
中国风，中国国潮插画风格

画质	基础设置
8K quality	--ar 9:16 --niji 5
8K 画质	画面比例 9:16　版本 Niji5

例 3. 夏至海报主图设计

绘画主体

Lots of lotus leaves floating on the lake, a girl dressed in white is on a boat
许多荷叶漂浮在湖面上，一个穿着白衣服的女孩在船上

场景

Summer, top view, wide angle shot, distant view, bright color, natural light, summer greens, dreamy
夏天，俯视图，广角拍摄，远景，明亮的色彩，自然光，夏天的绿色，梦幻

风格	画质	基础设置
Chinese style, Victorian style	8K quality	--ar 9:16 --niji 5
中国风，维多利亚风格	8K 画质	画面比例 9:16　版本 Niji5

第 6 章 海报主图设计

例 4. 秋分海报主图设计

绘画主体

Golden ginkgo trees, fallen leaves and pedestrians in the streets, autumn, landscape, falling leaves in the sky, vehicle

金黄的银杏树，街道上有落叶和行人，秋天，风景，空中有飘落的树叶，车辆

场景

Sharp focus, insanely detailed

锐利的焦点，疯狂的细节

画质

8K quality, high quality, high resolution

8K 画质，高品质，高分辨率

风格	基础设置
Ultra-realistic	--ar 2:3 --niji 5
超写实	画面比例 2:3
	版本 Niji5

153

例 5. 立冬海报主图设计

绘画主体

Snow, snowy cabins, plum trees, winter, natural light
雪，有积雪的小屋，梅花树，冬日，自然光

场景

Cold tone, white tone, a serene landscape, soft and dreamy atmosphere
冷色调，白色基调，宁静的风景，柔和梦幻的氛围

风格

Claborate-style painting, in the style of Chinese cultural themes, simple and elegant style
工笔画，以中国文化为主题的风格，简约典雅的风格

画质	基础设置
16K quality	--ar 2:3 --niji 5
16K 画质	画面比例 2:3　版本 Niji5

例6. 端午节海报主图设计

绘画主体

A cute little boy and girl are holding triangular zongzi wrapped in green leaves, zongzi is triangular and tower shape, zongzi is tied with yellow straw rope, zongzi has glutinous rice in it, highly detailed
一个可爱的小男孩和小女孩正抱着用绿色叶子包裹着的三角形粽子，粽子呈三角形和塔状，粽子用黄色的草绳系着，粽子里面有糯米，高细节

场景

The background is light green, green tones, sense of festival atmosphere
背景是淡绿色，绿调，节日气氛

风格

In the style of cute and dreamy, national tide style
可爱和梦幻的风格，国潮风格

画质

8K quality
8K 画质

基础设置

--ar 9:16 --niji 5
画面比例 9:16　版本 Niji5

例 7. 春节海报主图设计

绘画主体

Firework, lantern, Chinese courtyard, night, detailed character illustrations, crowd scenes
烟花，灯笼，中式庭院，夜晚，细致的人物插图，人群

场景

Chinese New Year background, red tone
中国新年背景，红色调

风格	画质	基础设置
Chinese style	8K quality	--ar 9:16 --niji 5
中国风	8K 画质	画面比例 9:16　版本 Niji5

例 8. 购物节海报主图设计

绘画主体
People are shopping, pushing shopping carts, many gift boxes and shopping bags
人们在购物，推着购物车，许多礼物盒和购物袋

场景
Promotional atmosphere, lively scene, ventilation scene, shopping bag elements, strong light, global illumination, front view
促销氛围，热闹场景，通风场景，购物袋元素，强光，全局照明，正视图

风格
Flat illustration, promotional poster design, vector illustration, bright colors
平面插画，促销海报设计，矢量图，明亮的色彩

画质
Ultra-detailed rendering, 8K quality
超细致渲染，8K 画质

基础设置
--ar 9:16 --v 5.2
画面比例 9:16　版本 V5.2

例 9. 父亲节海报主图设计

绘画主体
A father is holding his kid, smiling face
父亲抱着女儿，微笑的脸庞

场景
The background is blue sky and white clouds, warm, fresh
背景是蓝天白云，温馨，清新

风格
Cartoon, picture book style, hand-painted
卡通，绘本风格，手绘

画质	基础设置
8K quality	--ar 9:16 --niji 5
8K 画质	画面比例 9:16 版本 Niji5

Tips: 若生成的图像中人物占较大画面比例，遮挡了海报中文字的部分，可以先选择自己喜欢的画，在工具栏中单击"Zoom Out 2x"或"Zoom Out 1.5x"按钮，根据需求缩放 2 倍或 1.5 倍。

例 10. 母亲节海报主图设计

绘画主体

A cute girl with her mother, young mother looking down at her daughter, smiling face, mom's wearing an apron
一个可爱的女孩和她的妈妈，年轻的母亲正低头看着她的女儿，微笑的脸庞，妈妈穿着围裙

场景

In a lovely house, warm and bright, colorful, high quality, high saturation, outline light
在一个温馨的房子里，温暖明亮，色彩鲜艳，高品质，高饱和度，轮廓光

风格

Mother's day, cartoon
母亲节，卡通风格

画质

8K quality
8K 画质

基础设置

--ar 9:16 --niji 5
画面比例 9:16　版本 Niji5

例 11. 生日海报主图设计

绘画主体
Birthday cake, candles, in front of the window
生日蛋糕，蜡烛，摆在窗前

场景
Light gold, orange and pink, lovely atmosphere
淡金色的灯光，橙色和粉色色调，可爱的氛围

风格
Watercolor painting, in the style of anime arts
水彩画，动漫艺术的风格

画质	基础设置
HD	--ar 2:3 --niji 5
高清	画面比例 2:3　版本 Niji5

第 6 章 海报主图设计

例 12. 婚庆海报主图设计

绘画主体

A newly married couple, seaside wedding, by the beach, full body

一对新婚夫妻，海边婚礼，海滩边，全身

场景

Summer, lovely, romantically, rich colors, wedding, sunshine, flowers, pearls

夏天，可爱的，浪漫的，色彩丰富，婚礼，阳光，鲜花，珍珠

风格

Style of fairy tale book, wedding illustration poster

童话插画书风格，婚礼插画海报

画质	基础设置
8K quality	--ar 3:4 --niji 5
8K 画质	画面比例 3:4　版本 Niji5

161

例 13. 国潮海报主图设计

绘画主体

Chinese architecture, flowers, moon, rabbits, river
中式建筑，花，月亮，兔子，河

场景

Red tone, Song Dynasty, Mid-Autumn Festival atmosphere, warm color palette
红色调，宋朝，中秋节气氛，暖色调

风格

National tide style, Chinese style, package poster design
国潮风格，中国风，包装海报设计

画质

8K quality
8K 画质

基础设置

--ar 2:3 --niji 5
画面比例 2:3　版本 Niji5

例 14. 毕业季海报主图设计

绘画主体

Students in their graduation gowns, acceptance letter and flowers in hand, smiling face, highly detailed
身着毕业礼服的学生们，手里拿着录取通知书和鲜花，微笑的脸庞，高细节

场景

There are streamers in the sky, warm tone, depth of field, blue sky
空中有彩带，温暖的色调，景深，蓝天

风格

Cartoon
卡通风格

画质	基础设置
8K quality	--ar 9:16 --niji 5
8K 画质	画面比例 9:16　版本 Niji5

例15. 露营海报主图设计

绘画主体

Camping scene, people laughing at each other, pets, tent, picnic
露营场景，人们互相谈笑，宠物，帐篷，野餐

场景

Cheerful atmosphere, natural light, sunshine, flowers
欢乐的气氛，自然光，阳光，鲜花

风格	画质	基础设置
Flat illustration 平面插画	HD 高清	--ar 3:4 --niji 5 画面比例 3:4 版本 Niji5

例16. 旅游海报主图设计

绘画主体

The Great Wall, Shanghai, the London Eye
长城，上海，伦敦眼

场景

Wide angle, the color is mainly pastel blue purple and pink
广角，色调以淡蓝色、紫色、粉色为主

风格

Flat vector art illustration, travel poster
平面矢量艺术插图，旅游海报

画质	基础设置
8K quality 8K 画质	--ar 2:3 --v 5.2 画面比例 2:3　版本 V5.2

例 17. 饮料海报主图设计

绘画主体
Bottled drink, high quality, rich in details
瓶装饮料，高品质，细节丰富

场景
Glass bottle in seawater, tilted angle, bright and cheerful, high saturation colors, natural lighting, close-up shots, blue tint
海水中的玻璃瓶，倾斜角度，明亮欢快的色调，高饱和度色彩，自然采光，特写镜头，蓝色色调

风格
Adorable clay stop-motion animation, surrealism
可爱的黏土定格动画，超现实主义

画质
8K quality
8K 画质

基础设置
--ar 2:3 --niji 5
画面比例 2:3　版本 Niji5

例 18. 客房海报主图设计

绘画主体
Morning, sunrise, clouds, mountains, ultra-detailed, breakfast, outside the window
清晨，日出，云彩，山峦，超细节，早餐，窗外

场景
Romantic landscapes, large-scale paintings, heavy shading
浪漫的风景，大画幅，厚重的阴影

风格
In the style of fantasy illustrations
奇幻插画的风格

画质	基础设置
16K quality	--ar 3:2 --niji 5
16K 画质	画面比例 3:2　版本 Niji5

例 19. 护肤品海报主图设计

绘画主体
Skincare product, on the grass, delicate details
护肤品，放在草坪上，细节精致

场景
Clean background, flowers around, light and shadow, dreamy and warm atmosphere, central composition
干净的背景，周围有鲜花，光影，梦幻而温暖的氛围，中心构图

风格
Flat style, skincare illustration poster design
扁平风格，护肤插画海报设计

画质	基础设置
16K quality 16K 画质	--ar 3:2 --niji 5 画面比例 3:2　版本 Niji5

例 20. 口红海报主图设计

绘画主体
Lipstick, Chinese lipstick, exquisite gold wire details
口红，中式口红，精致的金丝细节

场景
Background is Chinese landscape, Chinese traditional waves, clouds and coral, plum blossom
背景为中国风景，中国传统的波浪，云彩和珊瑚，梅花

风格
Chinese style
中国风

画质
16K quality
16K 画质

基础设置
--ar 2:3 --v 5.2
画面比例 2:3　版本 V5.2

例 21. 香薰海报主图设计

绘画主体

Aromatherapy, scented candle, palm leaves, bright window light shadow
香薰，香薰蜡烛，棕榈叶，明亮的窗台光影

场景

Champagne tone, top angle, sill
香槟色调，顶部视角，窗台

风格

Products photoshoot, realistic
产品拍摄图，逼真

画质	基础设置
8K quality	--ar 2:3 --v 5.2
8K 画质	画面比例 2:3　版本 V5.2

例 22. 珠宝海报主图设计

绘画主体

Jewelry shooting, light colored stones, fine gloss, dry plants, earrings, bracelets, necklaces, rings
珠宝拍摄，浅色的宝石，光泽度好，干的植物，耳环，手链，项链，戒指

场景

Natural warm sunlight, light yellow background, soft atmosphere, high contrast accuracy, centered composition
自然暖光，淡黄色的背景，柔和的氛围，对比度高，居中构图

风格	画质	基础设置
Realistic photography	8K quality	--ar 2:3 --niji 5
写实摄影	8K 画质	画面比例 2:3　版本 Niji5

例 23. 教师节海报主图设计

绘画主体
A female teacher, the female teacher is giving a lecture to students, holding a textbook in her hand, natural light, students are listening attentively
一位女老师，这位女老师正在给学生讲课，手里拿着一本课本，自然光，学生们都在专心听讲

场景
A sunny day, a Chinese classroom
阳光明媚的一天，一间语文教室

风格
Asian Illustration style, the watercolor ink painting, elegant style
亚洲插画风格，水彩水墨画，典雅的风格

画质
HD
高清

基础设置
--ar 3:4 --niji 5
画面比例 3:4　版本 Niji5

例 24. 美食海报主图设计

绘画主体

Dessert with black tea, high quality, mouthwatering
甜点配红茶，高品质，令人垂涎欲滴的

场景

The background is a fancy restaurant, shallow depth of field, real colors and comfortable light
背景是一家高档餐厅，浅景深，真实的色彩和舒适的光线

风格

Food photography, ultra-realistic, photo realistic, photo realism, photorealistic
美食摄影，超写实，照片级写实，照片写实主义，逼真

画质

4K quality
4K 画质

基础设置

--ar 3:4 --v 5.2
画面比例 3:4　版本 V5.2

例 25. 茶具海报主图设计

绘画主体
Chinese teapot, green tea, tea cup, tea leaves
中国茶壶，绿茶，茶杯，茶叶

场景
A serene scene, incredibly detailed, sharpen, photography lighting
宁静的场景，令人难以置信的细节，锐化，摄影照明

风格
Cinematic shot, photos taken by Nikon, Chinese ink style, Chinese style
电影镜头，尼康拍摄的照片，中国水墨风，中国风

画质	基础设置
4K quality	--ar 2:3 --v 5.2
4K 画质	画面比例 2:3　版本 V5.2

例 26. 音乐节海报主图设计

绘画主体
Music festival, the band on the stage, the audience below, streamers were floating in the sky

音乐节，舞台上的乐队，下面的观众，天空中飘着彩带

场景
Light red and sky-blue, a vibrant atmosphere

淡红色的灯光和天蓝色色调，活力四射的氛围

风格
Concert advertisement, vibrant manga

演唱会广告，充满活力的漫画风格

画质
8K quality
8K 画质

基础设置
--ar 2:3 --v 5.2

画面比例 2:3　版本 V5.2

例 27. 运动会海报主图设计

绘画主体
Sports meeting, runners, racetrack, contest
运动会，运动员，跑道，比赛

场景
Vibrant airy scenes
充满活力的场景

风格
In the style of manga art, children's book illustrations
漫画风，童书插图

画质
HD
高清

基础设置
--ar 2:3 --niji 5
画面比例 2:3　版本 Niji5

例28. 家具海报主图设计

绘画主体
A living room, a sofa, furniture, sunlight, high-quality interior decoration
一间客厅，沙发，家具，日光，高品质的室内装饰

风格
In the style of soft shades, comfortable modern minimalism, natural minimalism
柔和色调的风格，宽敞舒适的现代极简主义，自然极简

画质
8K quality, high resolution
8K 画质，高分辨率

基础设置
--ar 2:3 --v 5.2
画面比例 2:3　版本 V5.2

> 例 29. 登山海报主图设计

绘画主体

Mount Qomolangma, a group of hikers, morning sun
珠穆朗玛峰，徒步旅行者，早晨的太阳

场景

Colorful, textured, patterned, fine art print
彩色，纹理，图案，精美的艺术印刷

风格

In the style of silk screening, vintage style, bold posters, delicate paper cut, art print
丝网印刷风格，复古风，大胆的海报，精致的剪纸，艺术印刷

画质　　　　　　　　基础设置
HD　　　　　　　　　--ar 2:3 --v 5.2
高清　　　　　　　　画面比例 2:3　版本 V5.2

例30. 空镜展台海报主图设计

绘画主体

Round marble podium, palm leaf, white flower, glowing round frame
圆形大理石讲台，棕榈叶，白色花朵，发光的圆形框架

场景

Abstract background, sunlight, concise and clear, studio lighting
抽象背景，阳光，简洁清晰，工作室照明

风格

Product display, the Chinese ink style painting, Chinese style
产品展示，中国水墨画，中国风

画质

3D rendering, high quality
三维渲染，高品质

基础设置

--ar 2:3 --v 5.2
画面比例 2:3　版本 V5.2

例 31. 耳机海报主图设计

绘画主体
Headphones
头戴式耳机

场景
Space background, planetary surface, product top light, saturation color, shadows, centered composition, light strip design
太空背景，星球表面，产品顶光，饱和配色，阴影，居中构图，光条设计

画质
8K quality
8K 画质

基础设置
--ar 3:4 --v 5.2
画面比例 3:4　版本 V5.2

例 32. 坤包海报主图设计

绘画主体
Luxury brand bag, embellish with flowers
奢侈品牌包，鲜花点缀

场景
Blue sky in the background, high-end atmosphere, simple luxury, bright scene
背景是蓝色的天空，高端氛围，简约奢华，明亮的场景

风格
Product photography, advertising photos, photorealistic
产品摄影，广告照片，逼真

画质
8K quality
8K 画质

基础设置
--ar 3:4 --v 5.2
画面比例 3:4　版本 V5.2

例 33. 香水海报主图设计

绘画主体

Perfume, perfume on the water, luxurious style, colorful light, ultra-detailed

香水，水上的香水，奢华感，多彩的光线，超细节

场景

Central composition

中心构图

风格

Product-view, magazine photography, photorealistic

产品视图，杂志风格，逼真

画质

8K quality

8K 画质

基础设置

--ar 2:3 --v 5.2

画面比例 2:3　版本 V5.2

例 34. 汽车海报主图设计

绘画主体
A brand new limousine
一辆崭新的高级轿车

风格
In the style of dynamic symmetry, Neo-Academism
动态对称的风格，新学院主义

场景
Mountains in the background, mirror presentation, light gray
背景是山脉，镜面呈现，浅灰色色调

画质
16K quality
16K 画质

基础设置
--ar 2:3 --v 5.2
画面比例 2:3　版本 V5.2

例1. 参考图生成定制头像设计

注：此图由 AI 生成

原始图片

绘画主体

A girl, looking sideways at the audience, with long seaweed hair
一个女孩，侧脸凝视观众，海藻般的长发

场景

Romantic caricature portraits, empty scenes, ultra-detailed, soft lighting, artistic, divine cinematic rim lighting
浪漫的漫画肖像，空旷的场景，超细节，柔和的灯光，艺术的，神圣的电影边缘照明

风格	画质	基础设置
Japanese manga style	8K quality	--iw 1.5 --niji 5
日本漫画风格	8K 画质	与图像相似权重 1.5　版本 Niji5

例 2. 参考图生成情侣头像设计

绘画主体
Chinese couple portrait, couple photo style, beautiful girl, handsome boy
中国情侣肖像，情侣照风格，美丽的女孩，帅气的男孩

场景
Cartoon, animation dreamworks, clean background, happy atmosphere
卡通，动画梦工厂，干净的背景，欢乐的气氛

风格
Animation dreamworks animation style, cartoon style
动画梦工厂动画风格，卡通风格

画质
3D rendering, 4K quality
三维渲染，4K 画质

基础设置
--v 5.2
版本 V5.2

注：此图由 AI 生成
原始图片

例3. 美妆参考设计

绘画主体
Asian girl has blue makeup and flowers on her face, exquisite makeup, creative makeup, glowing, dotted with sequins, mixed flowers and butterflies, simple mysterious lines on her face
亚洲女孩脸上有蓝色的妆容和花朵，精致的妆容，创意妆容，发光，点缀着亮片，混合了花朵和蝴蝶，脸上有简单神秘的线条

场景
Facial close-up, global illumination, eye-catching compositions
面部特写，全局照明，引人注目的构图

风格　　　　　　　　　**基础设置**
Naturalistic aesthetics　　　--v 5.2
自然主义美学　　　　　　　版本 V5.2

画质
Real photography, 8K quality
真实摄影，8K 画质

例 4. 表情包设计

绘画主体
Little girl expression stickers, various expressions, exaggerated expressions
小女孩表情贴纸，多种表情，夸张表情

场景
Cute, light background, cartoon, simple drawing form, tiles
可爱，浅色背景，卡通，简单绘画形式，平铺图

风格
Flat illustration style, cute style
平面插画风格，可爱风格

画质
HD
高清

基础设置
--v 6.0
版本 V6.0

例 5. 直播礼物特效设计

绘画主体
Rocket 3D icon, cartoon color, dark background, clay material
火箭三维图标，卡通色彩，深色背景，黏土材料

场景
Isometric, vibrant color scheme, shiny, neon lights, best details
等距，充满活力的配色方案，闪亮，霓虹灯，最佳细节

风格
Cyberpunk style
赛博朋克风格

画质
HD, 3D rendering, high resolution
高清，三维渲染，高分辨率

基础设置
--v 5.2
版本 V5.2

例 6. 灯牌设计

绘画主体

Acrylic light sign design, music theme, musical notes and microphone elements, pink and purple, glow, neon lignts

亚克力灯牌设计，音乐主题，音符和麦克风元素，粉色和紫色，发光，霓虹灯

场景

Simple lines, dark background, creative shapes, acrylic material, multiple perspectives

简单的线条，深色的背景，创意的形状，亚克力材料，多个视角

风格	画质	基础设置
Advertising design, plane	Detailed details, Ultra HD, 8K quality	--v 5.2
广告设计，平面	详细的细节，超高清，8K 画质	版本 V5.2

例 7.APP 开屏界面设计

绘画主体

Graphic design about shopping websites, shopping carts, clothing, electronic products and other elements
关于购物网站、购物车、服装、电子产品等元素的平面设计

场景

Low purity colors, harmonious combinations, central composition, ultra-modern, award-winning design
低纯度色彩，和谐组合，中心构图，超现代，获奖设计

风格

Simple style, graphic design, graphic illustration, ultra-futurism, web design
简约风格，平面设计，平面插画，超未来主义，网页设计

画质	基础设置
HD	--ar 9:16 --v 5.2
高清	画面比例 9:16　版本 V5.2

例 8. 弹窗广告设计

绘画主体
Web advertising, music website official LOGO, musical notes, musical instrument elements
网页广告，音乐网站官方标志，音符，乐器元素

场景
Vibrant and dynamic atmosphere, cute toy sculptures, dynamic scenes, vivid energy explosion
活力动感氛围，可爱玩具雕塑，动态场景，生动的能量爆炸

风格
Website promotion style
网站推广风格

画质
C4D rendering, 32K quality
C4D 渲染，32K 画质

基础设置
--v 5.2
版本 V5.2

例 9. 发布会主视觉设计

绘画主体
Main visual design, iceberg theme, surrounded by blue light chains
主视觉设计，冰山主题，蓝色光链包围

场景
Simple, wide-angle lens, technological atmosphere, Arctic atmosphere
简约的，广角镜头，科技氛围，北极氛围

风格	基础设置
Graphic design	--ar 16:9 --v 5.2
平面设计	画面比例 16:9　版本 V5.2

画质
Fine details, 3D rendering, Ultra HD, 32K quality
精细细节，三维渲染，超高清，32K 画质

例 10. 数据分析 PPT 模板设计

绘画主体

Flat design layout about consumption data analysis, light blue and white color, chart, text, mind map, ultra modern

关于消费数据分析的平面设计布局，浅蓝色和白色，图表，文本，思维导图，超现代

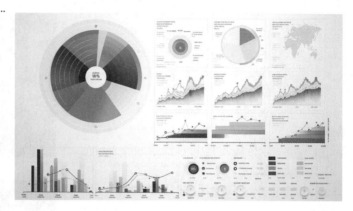

场景

Central composition, award-winning design

中心构图，获奖设计

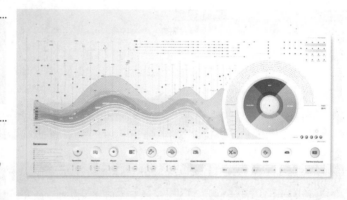

风格

Simple style, ultra-future, graphic design, web design

简约风格，超未来，平面设计，网页设计

画质

HD
高清

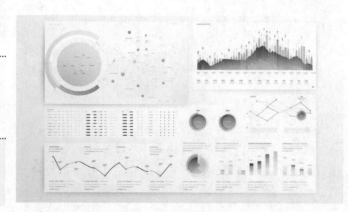

基础设置

--ar 16:9 --v 5.2
画面比例 16:9　版本 V5.2

例 11. 美妆直播间设计

绘画主体

The layout of the live broadcast room, there are makeup mirrors and cosmetics on the table, and cosmetics display cabinets on the background wall, mainly red and white

直播间布局,桌子上有化妆镜和化妆品,背景墙上有化妆品展示柜,以红白为主

场景

Natural theme atmosphere, subjective perspective, global illumination, central composition

自然主题氛围,主观视角,全局照明,中心构图

风格

Interior design style

室内设计风格

画质

Realistic photography, ultra-fine details, 32K image quality

写实摄影,超精细细节,32K 图像质量

基础设置

--ar 3:5 --v 5.2

画面比例 3:5 版本 V5.2

例 12. 夏日氛围直播间背景设计

绘画主体

Summer market, outdoor umbrellas, various fruits, beach, ocean
夏季集市，户外遮阳伞，各种水果，海滩，海洋

场景

Summer atmosphere, high saturation colors, harmonious combination, natural and healthy atmosphere
夏日氛围，高饱和度色彩，和谐组合，自然健康的氛围

风格

Outdoor set design, studio set design
户外布景设计，工作室布景设计

画质

C4D rendering, UHD, high resolution, 8K quality
C4D 渲染，超高清，高分辨率，8K 画质

基础设置

--ar 3: 5 --v 5.2
画面比例 3: 5　版本 V5.2

例 13. 梦幻氛围光感直播间背景设计

绘画主体

Psychedelic dream style studio layout, blue, purple and orange lights
迷幻梦幻风格的工作室布局，蓝紫橙灯光

场景

High saturation colors, cyberpunk atmosphere, asymmetrical composition, front view, low-fidelity aesthetics, harmony
高饱和度色彩，赛博朋克氛围，不对称构图，正视图，低保真美学，和谐

风格	画质	基础设置
Studio set 工作室布景	Ultra HD, high resolution, 8K quality 超高清，高分辨率，8K 画质	--ar 3：5 --v 5.2 画面比例 3：5　版本 V5.2

例 14. 科技感氛围直播间背景设计

绘画主体
The live broadcast background wall of the space starship cockpit, is mainly orange and white, with a curved surface
宇宙飞船驾驶舱直播背景墙，以橙白色为主，曲面

场景
Soft colors, technological atmosphere, leaving room for imagination, studio lighting
色彩柔和，科技氛围，留有想象空间，工作室照明

风格
Interior design, studio set, futuristic sci-fi, minimalist
室内设计，工作室布景，未来科幻，简约

画质	基础设置
Ultra HD, 8K quality 超高清，8K 画质	--ar 9:16 --v 5.2 画面比例 9:16　版本 V5.2

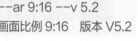

例 15. 中国风直播间背景设计

绘画主体

Minimalist stage design, small stage, Chinese architecture and bamboo forest, moon
极简舞台设计，小舞台，中式建筑和竹林，月亮

场景

Stage focus, simple brushstrokes, more white space, natural and rustic atmosphere, studio lighting, high quality, ultra-detailed, quiet and harmonious style, center of composition, meticulous texture and craftsmanship
舞台焦点，简单的笔触，更多的留白，自然质朴的氛围，工作室照明，高品质，超细节，安静和谐的风格，构图中心，细致的纹理和工艺

风格

Graphic art, natural art
图形艺术，自然艺术

画质

Ultra HD, 8K quality
超高清，8K 画质

基础设置

--ar 9:16 --v 5.2
画面比例 9:16 版本 V5.2

例 16. 新年氛围直播间背景设计

绘画主体

The New Year atmosphere live broadcast background wall mainly includes red, gold and white, lanterns, firecrackers, wooden furniture and other elements
新年气氛直播背景墙，主要有红色，金白色，灯笼，鞭炮和木制家具等元素

场景

Studio lighting, New Year atmosphere
工作室照明，新年氛围

风格

Chinese style, studio set
中国风，工作室布景

画质

Ultra HD, 8K quality
超高清，8K 画质

基础设置

--ar 9:16 --v 5.2
画面比例 9:16　版本 V5.2

例1. 微距摄影效果

绘画主体
Macro photography, growing moss on rocks, green, fresh
微距摄影,岩石上生长着苔藓,绿色,清新

场景
Miniature core, blurred background, extreme close-up view, exaggerated proportions
微型核心,模糊的背景,极端特写视图,夸张的比例

风格
Visually striking, realistic, elegant lines
具有视觉冲击力,逼真,优雅的线条

画质
32K UHD
32K 超高清

基础设置
--ar 16:10 --v 5.2
画面比例 16:10 版本 V5.2

例 2. 鱼眼镜头效果

绘画主体

The GoPro view of Panda in armor walking down the street
GoPro 拍摄的一只穿盔甲的熊猫走在街上的画面

场景

Fish-eye effect, walking in the bustling street, city night views in the background
鱼眼效果，走在繁华的街道上，背景是城市夜景

风格	画质	基础设置
Fine art cinematic portrait photography, impressive, street art sensibilities 美术电影肖像摄影，令人印象深刻，街头艺术感	Realist details 现实细节	--ar 16:9 --v 5.2 画面比例 16:9　版本 V5.2

例 3. 广角镜头效果

绘画主体

A wooden bridge, with snow cover, realistic blue skies, snow scenes
一座木桥，有积雪覆盖，逼真的蓝天，雪景

场景

Strong light effect, ultra wide shot, extreme angle
强光效，超广角，极限角度

风格

Photography by Evgeni Dinev
叶夫根尼·迪涅夫摄

画质

32K UHD
32K 超高清

基础设置

--ar 16:9 --v 5.2
画面比例 16:9　版本 V5.2

例 4. 魔幻现实主义摄影效果

绘画主体
A little boy stands in front of a giant piranha, in the tropical rainforest
一个小男孩站在巨型食人鱼前，在热带雨林中

场景
Sunny weather, asymmetrical composition
阳光明媚的天气，不对称构图

风格
1999s, realistic photography
20 世纪 90 年代风格，写实摄影

基础设置
--ar 5:3 --v 5.2
画面比例 5:3　版本 V5.2

例 5. 极光照片效果

绘画主体
The aurora borealis as it lights up a hut on a snowy landscape, light green and sky-blue
北极光照亮了雪地上的小屋，浅绿色和天蓝色

场景
Big panorama, photo-realistic landscapes, ethereal images, capturing moments
大全景，照片级的写实风景，空灵的图像，捕捉瞬间

风格
Post-minimalist structures, romantic and beautiful
后极简主义结构，浪漫又美丽

画质
HD
高清

基础设置
--ar 16:9 --v 5.2
画面比例 16:9　版本 V5.2

例 6. 动物类摄影效果

绘画主体

A red flamingo in water looking for food, light red and dark pink, mysterious jungle, photo-realistic landscapes

水中寻找食物的红色火烈鸟,浅红色和深粉色,神秘的丛林,照片级的写实风景

场景

Center the composition, precisionist, luminous reflections

构图居中,精准,发光反射

风格

Intense coloration, realistic animal portraits, live action shooting

浓烈的色彩,逼真的动物肖像,现场实拍

画质	基础设置
3840*2160	--ar 16:9 --v 5.2
3840*2160	画面比例 16:9　版本 V5.2

例 7. 花卉类摄影效果

绘画主体
A flower is in full bloom among some plants
一些植物中盛开着一朵花

场景
Against the sunset, quiet and charming, soft lights, low angle view
夕阳映衬，静谧迷人，灯光柔和，低视角

风格
In the style of ZEISS batis 18mm f/2.8, biomorphic forms, figurative naturalism, minimalist photography, perfect details
蔡司 batis 18mm f/2.8 风格，生物形态，具象自然主义，极简摄影，细节完美

画质
High resolution
高分辨率

基础设置
--ar 16:9 --v 5.2
画面比例 16:9　版本 V5.2

例 8. 无人机俯拍效果

绘画主体
A small island with blue water around in the ocean
海洋中有一个周围是蓝色海水的小岛

场景
Aerial view, drone overhead shooting
鸟瞰图，无人机俯拍

风格
One-of-a-kind pieces, Sumatraism, Millennial Aesthetics, imposing monumentality
独一无二的作品，苏门答腊主义，千禧年美学，庄严宏伟

画质
32K UHD
32K 超高清

基础设置
--ar 16:9 --v 5.2
画面比例 16:9 版本 V5.2

例 9. 红外摄影效果

绘画主体

An infrared photographic image, a big tree on a grassy hillside
红外摄影图像，长满青草的山坡上的一棵大树

场景

Natural lighting, asymmetrical composition, spectacular backgrounds, realistic and romantic scenery
自然采光，不对称构图，壮观的背景，真实又浪漫的风景

风格

Infrared photography, photo-realistic techniques
红外摄影，照片级写实技巧

基础设置

--ar 16:9 --v 5.2
画面比例 16:9　版本 V5.2

例 10. 光绘效果

绘画主体

Light drawing renderings of the circle, light painting, dark orange and light gold, spirals

圆形光绘效果图，光绘画，深橙色和浅金色，螺旋图案

场景

Rule of thirds composition

三分法构图

风格

Colorful explosions, splash the sparks, night photography

色彩缤纷的爆炸，飞溅的火花，夜间摄影

画质

Hyper quality

超高品质

基础设置

--ar 5:3 --v 5.2

画面比例 5:3　版本 V5.2

例 11. X 射线摄影效果

绘画主体
X-ray effect, rose and leaves
X 射线效果，玫瑰和叶子

场景
Realistic portrayal of light and shadow, photorealistic accuracy, projection mapping, edge transparency
真实的光影刻画，逼真精度，投影映射，边缘透明

风格
Flat, dark white and light silver, meticulous realism, ethereal photograms
平面，深白和浅银，细致写实，空灵照片

基础设置
--ar 5:3 --v 5.2
画面比例 5:3　版本 V5.2

例12. 双重曝光效果

绘画主体

Double exposure of a beautiful girl and flowers, captivating gaze, profile, light gray and amber, dreamy collages

美丽女孩和花朵的双重曝光,迷人的目光,侧面轮廓,浅灰色和琥珀色,梦幻的拼贴画

场景

Layered landscapes, captures the essence of nature, psychological landscapes

层次分明的风景,捕捉自然的本质,心理景观

风格

Photograph by Christoffer Relander, double exposure photography

克里斯托弗·雷兰德摄,双重曝光摄影

基础设置

--ar 2:3 --v 5.2

画面比例 2:3　版本 V5.2

例 13. 人物情绪摄影效果

绘画主体
A young lady with long hair is looking out of the window, depressed, heavy makeup, side part, beautiful, elegant, emotive faces
一位长发女士望着窗外，忧郁，浓妆艳抹，侧分，美丽，优雅，动情的脸庞

场景
Dramatic chiaroscuro effects, facial close-up, full of atmosphere
戏剧性的明暗对比效果，面部特写，气氛十足

风格
Distressed edges, individually distinct, strong emotional impact, photograph by Deni Pesto, character emotional photography
仿旧边缘，个性鲜明，情感冲击力强，丹尼·佩斯托摄，人物情感摄影

基础设置
--ar 5:3 --v 5.2
画面比例 5:3　版本 V5.2

例 14. 展示台摄影效果

绘画主体

Round marble booth, product display stand, simple irregular metal luminous frame
圆形大理石展台，产品展示架，简洁的不规则金属发光框架

场景

Undersea background, splashing water drops, soft light, central composition, simple and clear, studio lighting
海底背景，飞溅的水珠，柔和光线，中心构图，简洁清晰，工作室照明

风格	画质	基础设置
Realistic, photography 写实摄影	3D rendering, detailed details, high quality rendering, 8K quality 三维渲染，详细的细节，高品质渲染，8K 画质	--ar 3:4 --v 5.2 画面比例 3:4 版本 V5.2

例 15. 人物景别摄影效果

绘画主体
A woman dressed in traditional Chinese red clothing standing in the snow, exquisite ancient Chinese hair bun
一个穿着中国传统红衣的女人站在雪地里，精致的中国古代发髻

场景
Medium shot, 35mm lens
中景，35mm 镜头

风格
Ancient Chinese art style, charming photo album, cute and dreamy, mythical themes, photograph by Imogen Cunningham
中国古代艺术风格，迷人相册，可爱梦幻，神话主题，伊莫金·坎宁安摄

画质
Ultra-high resolution
超高分辨率

基础设置
--ar 5:3 --v 5.2
画面比例 5:3　版本 V5.2

例 16. 色彩焦点照片效果

绘画主体

A little red leaf lay down on cement in a shadow, complete, beautiful leaf
一片小红叶躺在水泥地上的阴影里，完整、美丽的叶子

场景

Diagonal composition, low exposure, strong contrast of light and shadow, focus on the leaf, juxtaposition of light and shadow, chiaroscuro
对角线构图，低曝光，光影对比强烈，以树叶为中心，光影并置，明暗对比

风格	画质	基础设置
Color focus style, raw street photography, matte photo 色彩聚焦风格，原始街头摄影，哑光照片	UHD image 超高清图像	--ar 9:16 --v 5.2 画面比例 9:16 版本 V5.2

例 17. 散景效果

绘画主体
Bokeh effect, a street Christmas tree, dark amber and gold, winter night, faint spot of light

散景效果,街上的一棵圣诞树,深琥珀色和金色,冬夜,微弱的光点

场景
Bokeh, close-up shots, shot on 18mm, f/5.6, festive atmosphere

背景虚化,特写镜头,18mm 拍摄,f/5.6,节日氛围

画质
8K resolution

8K 分辨率

基础设置
--ar 5:3 --v 5.2

画面比例 5:3 版本 V5.2

例 18. 动态模糊效果

绘画主体

A girl is running, long wavy hair, side profile, 1990s, hurried, anxious, disoriented, running, city streets at night

一个女孩正在奔跑，长卷发，侧面轮廓，20世纪90年代，匆忙，焦虑，迷茫，奔跑，夜晚的城市街道

场景

Shot on 35mm lens, face shot, mixed light, neon lights, slow shutter, motion blur

35mm 镜头拍摄，脸部拍摄，混合光，霓虹灯，慢快门，运动模糊

风格

Wong Kar-wai style, photograph by Christopher Doyle, last century style, classic retro, matte, pictorial film

王家卫风格，克里斯托弗·道尔摄，上世纪风格，经典复古，哑光，写真胶片

基础设置

--ar 2:3 --v 5.2

画面比例 2:3　版本 V5.2

例 19. 移轴摄影效果

绘画主体

A beautiful view of small white buildings, Santorini, blue sea behind
白色小建筑的美丽景色，圣托里尼岛，后面是蓝色的大海

场景

Tilt-shift photography, low light, selective focus, asymmetrical composition, scenery shot, sun-soaked colours
移轴摄影，弱光，选择性对焦，不对称构图，风景拍摄，阳光浸透的颜色

风格

Mediterranean architecture, architectural chic
地中海建筑，别致的建筑

画质	基础设置
Realistic details 逼真的细节	--ar 16:9 --v 5.2 画面比例 16:9　版本 V5.2

例 20. 轮廓光摄影效果

绘画主体

A woman is silhouetted against the dark, gentle expressions, distinctive nose, delicate features
黑暗中映衬出一个女人的轮廓，温柔的表情，特征鲜明的鼻子，精致的五官

场景

Contour light, tamron 24mm f/2.8, silhouette lighting, side view, symmetrical composition
轮廓光，腾龙 24mm f/2.8，剪影光，侧视图，对称构图

风格

Caravaggesque chiaroscuro, monochromatic, strong contrast between light and dark, feminine portraiture, magazine cover, extreme details
卡拉瓦乔风格的明暗对比，单色，强烈的明暗对比，女性肖像，杂志封面，极致的细节

基础设置

--ar 3:4 --v 5.2
画面比例 3:4   版本 V5.2

例 21. 飞溅摄影效果

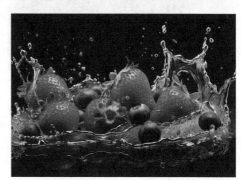

绘画主体

The moment a strawberry and blueberry collide with water, with water splashing, tumblewave, splash moment
草莓和蓝莓与水的碰撞瞬间，水花四溅，翻腾，溅起水花的瞬间

场景

Shot on 70mm, black background, rich layers and complex composition
70mm 拍摄，黑色背景，层次丰富，构图复杂

风格

Commercial style, vivid colors, captured essence of the moment, dynamic and action-packed
商业风格，色彩鲜艳，捕捉瞬间精髓，动感十足

画质

8K quality
8K 画质

基础设置

--ar 16:9 --v 5.2
画面比例 16:9　版本 V5.2

例 22. 长焦摄影效果

绘画主体

Hummingbird is flying towards a pink flower, as if in front of my eyes, characterized mouth
蜂鸟正飞向一朵粉色的花朵，仿佛就在我的眼前，具有特征的嘴

场景

Telephoto lens, blurred background
长焦镜头，背景虚化

风格

Meticulous technique, animated GIF, nature's wonder, captivating
细致的技法，GIF 动画，大自然的鬼斧神工，令人着迷

画质

16K resolution
16K 分辨率

基础设置

--ar 5:3 --v 5.2
画面比例 5:3　版本 V5.2

例 23. 特写摄影效果

绘画主体

A close-up view of an owl's eye, with orange colored eyes, single eye
猫头鹰眼睛的特写视图，橙色的眼睛，单眼

场景

Macro zoom, local close-up, realistic light and shadow
微距变焦，局部特写，逼真光影

风格

Photo by hasselblad 500c/m, hyper-realistic animal illustrations, accurate bird specimens, vivid, full vibrant, highly detailed
哈苏 500c/m 拍摄，超写实的动物插图，准确的鸟类标本，生动，充满活力，细节丰富

画质
UHD image
超高清图像

基础设置
--ar 5:3 --v 5.2
画面比例 5:3　版本 V5.2

例 24. 复古艺术照摄影效果

绘画主体

A beautiful woman sitting in front of a mirror, wearing an Hepburn style skirt, elegant sitting posture, 1950s, black and white

一位美女坐在镜子前，穿着赫本风格的裙子，优雅的坐姿，20 世纪 50 年代，黑白

场景

Contrasting light and dark tones, sharp focus, cinematic mood

明暗对比，焦点清晰，电影氛围

风格

Black and white photo, vintage effect, photo by Kodak, art deco-inspired, classic style, movie still, romantic emotion

黑白照片，复古效果，柯达拍摄，装饰艺术风格，经典风格，电影剧照，浪漫情感

基础设置

--ar 2:3 --v 5.2

画面比例 2:3 版本 V5.2

例 25. 封面人物摄影效果

绘画主体

Poster, mysterious backdrops, red and black, elegant clothing, graceful curves
海报，神秘的背景，红黑相间，优雅的服饰，曼妙的曲线

场景

Soft focal points, close shot
柔焦，近景

风格

Poster art, shadowy drama, dramatic use of color, photograph by Annie Leibovitz
海报艺术，阴暗戏剧，色彩的戏剧性运用，安妮·莱博维茨摄

画质

32K UHD
32K 超高清

基础设置

--ar 3:5 --v 5.2
画面比例 3:5　版本 V5.2

例 26. 婚纱摄影效果

绘画主体

Wedding photos, a married couple sitting on the crooked moon at a water, light silver and gold
婚纱照，一对新人坐在水边的弯月上，浅银和金色

场景

Exquisite theatrical lighting, medium shot, photo by Canon 5D Mark Ⅲ
精美的剧场灯光，中景，佳能 5DMark Ⅲ 拍摄

风格

Simple and elegant style, eye-catching, enchanting, glittery, dreamy, essence of the moment
简约大方的风格，夺目，妖娆，闪亮，梦幻，抓住瞬间的精髓

画质

8K resolution
8K 分辨率

基础设置

--ar 5:3 --v 5.2
画面比例 5:3　版本 V5.2

例 27. 工作照摄影效果

绘画主体

Chinese professional female, executives, office professional suit, with confident smile, crossed arms, long hair
中国职业女性，高管，办公室职业套装，自信的微笑，交叉双臂，长发

场景

Office background, studio lighting, bokeh, half body shot, photo by ZEISS
办公室背景，工作室照明，背景虚化，半身照，蔡司拍摄

画质

Super details, 4K quality, best quality
超级细节，4K 画质，最佳质量

基础设置

--ar 3:4 --v 5.2
画面比例 3:4　版本 V5.2

例 28. 食品素材摄影效果

绘画主体
On the wooden table are beef, vegetables and beans
木桌上摆着牛肉、蔬菜和豆类

场景
Soft focal points, aerial view, clean background
柔焦，鸟瞰图，干净的背景

风格
Bold and vibrant, realistic, appetizing, rich and neatly arranged
大胆而充满活力，写实，诱人的，丰富而整齐

画质
HD
高清

基础设置
--ar 16:9 --v 5.2
画面比例 16:9　版本 V5.2

例 29. 美食摄影效果

绘画主体
A portion of spaghetti on the table
桌上的一份意大利面

场景
Medium shot, moderate saturation, vivid contrast, depth of field, shot on 85mm, f/ 2.8
中景，饱和度适中，对比度鲜明，景深，85mm 镜头拍摄，f/ 2.8 光圈拍摄

风格
Delicious and tempting
美味诱人

画质
Best quality
最好的质量

基础设置
--ar 5:3 --v 5.2
画面比例 5:3　版本 V5.2

例 30. 烤箱内摄影效果

绘画主体
The egg tarts are baking in the oven, golden tart, warm yellow light in the oven
蛋挞正在烤箱里烘烤，金黄的蛋挞，烤箱里有暖黄的灯光

场景
Medium shot, high angle shot, focus on the egg tarts, outside oven view
中景，高角度，聚焦蛋挞，从烤箱外部看

风格
Delicious and crispy, line up, food photography, physical photography
香脆可口，排队，美食摄影，实物摄影

画质
UHD
超高清

基础设置
--ar 5:3 --v 5.2
画面比例 5:3　版本 V5.2

第 9 章
装帧及文创设计

例1. 画册设计

绘画主体

Picture book production, book binding design, picture albums
画册制作，书籍装帧设计，画册

场景	画质	基础设置
Clean background	HD	--v 5.2
干净的背景	高清	版本 V5.2

例2. 布艺封面设计

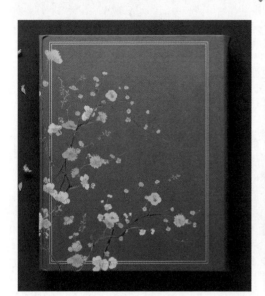

绘画主体
Book binding design, a book
书籍装帧设计，一本书

场景
The cover is made of dark blue fabric, Chinese flowers
封面是由深蓝色布料做成的，中式花朵

风格
Literature and art, the feeling of sackcloth
文学和艺术，麻布质感

画质	基础设置
HD	--v 5.2
高清	版本 V5.2

例3. 地理书籍封面设计

绘画主体
Geographic magazine, book binding design, a book
地理杂志，书籍装帧设计，一本书

场景
The cover shows mountains and rivers, geography
封面是山川河流，地理

画质	基础设置
HD 高清	--v 5.2 版本 V5.2

例4. 儿童书籍封面设计

绘画主体
Children's books, book binding design, a book
儿童书籍，书籍装帧设计，一本书

场景
Bright colors, fresh atmosphere
明亮的色彩，清新的氛围

风格	画质	基础设置
Watercolor style 水彩风格	HD 高清	--v 5.2 版本 V5.2

例 5. 科幻书籍封面设计

绘画主体
Science fiction books, book binding design, a book
科幻书籍，书籍装帧设计，一本书

场景
The cover shows tall buildings of the future
封面是未来的高楼大厦

风格	画质	基础设置
Cyberpunk style	HD	--v 5.2
赛博朋克风格	高清	版本 V5.2

例 6. 时尚杂志封面设计

绘画主体

A magazine, book binding design
一本杂志，书籍装帧设计

场景

A fashion magazine cover
一本时尚杂志的封面

风格	画质	基础设置
Sense of art	HD	--v 5.2
艺术感	高清	版本 V5.2

例 7. 精装书封面设计

绘画主体
A book, book binding design
一本书，书籍装帧设计

场景
Hard cover, clean background
硬壳封面，干净的背景

画质	基础设置
HD	--v 5.2
高清	版本 V5.2

例 8. 古风风格书籍封面设计

绘画主体
A book, book binding design, Chinese poetry, Chinese style book design
一本书，书籍装帧设计，中国诗词，中国风书籍设计

场景
Clean background
干净的背景

风格	画质	基础设置
Chinese style	HD	--v 5.2
中国风	高清	版本 V5.2

例 9. 镂空封面设计

绘画主体
A book, book binding design
一本书,书籍装帧设计

场景
Cut-out cover, clean background
镂空封面,干净的背景

画质
HD
高清

基础设置
--v 5.2
版本 V5.2

例 10. 说明书排版设计

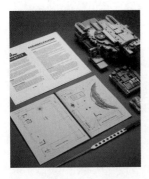

绘画主体
Product specification design
产品说明书设计

场景
The instruction sheet is on the table
说明书在桌子上

画质	基础设置
HD	--v 5.2
高清	版本 V5.2

例 11. 定制信封设计

绘画主体
Envelope design
信封设计

场景	风格
Clean background	Simple style
干净的背景	简约风格

画质	基础设置
HD	--v 5.2
高清	版本 V5.2

例 12. 线圈本设计

绘画主体

Coil notebook, notebook, notebook cover, notebook design
线圈笔记本，笔记本，笔记本封面，笔记本设计

场景

Hard cover, the cover features trees
硬壳封面，封面中有树的设计

风格

Simple style, watercolour style
简约风格，水彩风格

画质	基础设置
HD	--v 5.2
高清	版本 V5.2

例 13. 邀请函设计

绘画主体

Blue ribbon business invitation, table, gold lace, solid color, dark blue gold
蓝丝带商务邀请函，桌子，金色蕾丝，纯色，深蓝金色

场景

Bright light, luster
光线明亮，光泽感

风格	画质	基础设置
Retro style	HD	--v 6.0
复古风格	高清	版本 V6.0

例 14. 立体贺卡设计

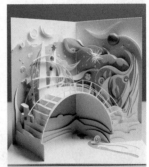

绘画主体
Three-dimensional greeting card design, open card, card design, greeting card
立体贺卡设计，翻开的贺卡，贺卡设计，贺卡

场景
Clean background
干净的背景

画质	基础设置
HD	--v 5.2
高清	版本 V5.2

例 15. 线装书工艺设计

绘画主体
A book, book binding design, books bound with thread, thread-bound book
一本书，书籍装帧设计，用线装订的书，线装书

场景
Clean background
干净的背景

画质	基础设置
HD	--v 5.2
高清	版本 V5.2

例 16. 折页宣传册设计

绘画主体
Brochure design, folding, pamphlet
宣传册设计，折页，小册子

场景
Clean background
干净的背景

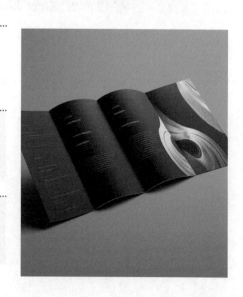

画质	基础设置
HD	--niji 5
高清	版本 Niji5

例 17. 立体书设计

绘画主体
Three-dimensional book, book binding design
立体书，书籍装帧设计

场景
Chinese landscape art, high quality material, space relief painting, standing on paper, open pages, cool tone
中国山水艺术，高品质的材质，空间浮雕画，立在纸上，翻开的书页，冷色调

风格
Chinese style, ink style painting, wash painting, minimalism
中国风，水墨画风格，洗染画，极简主义

画质
HD
高清

基础设置
--v 5.2
版本 V5.2

例 18. 书脊陈列设计

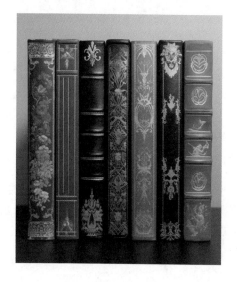

绘画主体
Book spine design
书籍书脊设计

场景
Solid background
纯色背景

画质	基础设置
HD	--v 5.2
高清	版本 V5.2

例 19. 明信片设计

绘画主体
Postcard, typography design,
set of flat vector illustrations
明信片，排版设计，一组平面矢量插图

场景
Nature and landscape of Europe
欧洲的自然美观

风格	画质	基础设置
Flat style	HD	--niji 5
扁平风格	高清	版本 Niji5

例 20. 名片设计

绘画主体
Name card, name card design
名片，名片设计

风格
Simple style
简约风格

画质	基础设置
HD	--v 5.2
高清	版本 V5.2

例 21. 伴手礼设计

绘画主体
Package design, gift box design
包装设计，礼品盒设计

场景	风格
Clean background	Chinese style
干净的背景	中式风格

画质	基础设置
HD	--v 5.2
高清	版本 V5.2

例 22. 便签设计

绘画主体
Memo notepaper, memo
便签纸，便签

场景
A memo on the table, Chinese style pattern, the memo have a decorative pattern
桌上有便签，中国风花纹，便签上有装饰花纹

风格
Chinese style
中国风

画质
16K quality
16K 画质

基础设置
--v 5.2
版本 V5.2

例 23. 特殊工艺贺卡设计

绘画主体
Greeting card design,
New Year greeting card
贺卡设计，新年贺卡

场景	风格
Red tone	Chinese style, paper carving art
红色调	中国风，纸雕艺术

画质
HD
高清

基础设置
--v 5.2
版本 V5.2

例 24. 词典设计

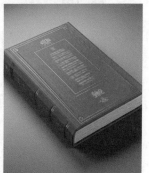

绘画主体

English-Chinese dictionary, translation dictionary, dictionary design
英汉词典，翻译词典，词典设计

场景

Clean background
干净的背景

画质	基础设置
HD	--v 5.2
高清	版本 V5.2

例 25. 相册排版设计

绘画主体

Photo album, photo, collections of photographs
相册，照片，收藏的照片

场景

On the table
放在桌子上

风格

Retro tone
复古风

画质

HD
高清

基础设置

--v 5.2
版本 V5.2

例 26. 烫金工艺封面设计

绘画主体
Book binding design, a book
书籍装帧设计,一本书

场景
The cover is made of thick cardboard, bronzing process, there is wheat and rice on the picture
封面用厚纸板做成,烫金工艺,画面上有小麦和大米

风格	画质	基础设置
Literature and art, ancient	HD	--v 5.2
文学和艺术,复古风	高清	版本 V5.2

例 27. 定制钢笔设计

绘画主体
Fountain pen, pen, Chinese style pattern
钢笔，笔，中国风花纹

场景
Pen put on the desk
钢笔放在桌子上

风格
Chinese style
中国风

画质
8K quality
8K 画质

基础设置
--ar 3:2 --v 5.2
画面比例 3:2　版本 V5.2

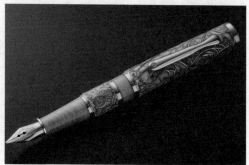

例 28. 邮票设计

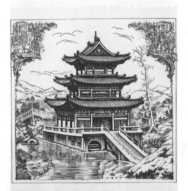

绘画主体
Vintage postage stamp
复古邮票

场景
Chinese architecture, line engraving
中国建筑，线条雕刻

风格	画质	基础设置
Ink painting style	HD	--v 5.2
水墨画风格	高清	版本 V5.2

例 29. VIP 卡设计

绘画主体
Membership card design
会员卡制作

场景
Solid color background, printed design
纯色背景，印花设计

画质
HD
高清

基础设置
--v 5.2
版本 V5.2

例 30. CD 封套设计

绘画主体

Music album, physical album, music CD, tape
音乐专辑，实体专辑，音乐 CD，磁带

场景

The black music disc was placed in a music album box with a cover of summer flowers, three-dimensional
黑色的唱片被放在一个封面是夏日鲜花的唱片盒里，立体感

风格

Artistic sense
艺术感

画质

HD
高清

基础设置

--v 5.2
版本 V5.2

例 31. 手账版式设计

绘画主体	场景
Techo, bullet journal cover, book binding design 手账，手账封面，书籍装帧设计	On the table 放在桌子上

画质	基础设置
HD 高清	--v 5.2 版本 V5.2

第 10 章

展品设计

例1. 旗袍橱窗模特设计

绘画主体
Window mannequin wearing olive green cheongsam, long skirt, noble silk
穿橄榄绿旗袍的橱窗模特，长裙，高贵的丝绸

场景
Smooth tailoring, expensive and luxurious, elegant and luxurious, simple design, dreamy texture, ultra-fine details, front view, through glass perspective
流畅的剪裁，昂贵奢华，优雅华贵，简约的设计，梦幻般的质感，超精细的细节，正视图，透过玻璃透视

风格
Simple Chinese design
简约的中式设计

画质	基础设置
Real photography, 32K quality 真实摄影，32K 画质	--v 5.2 版本 V 5.2

例 2. 运动服装橱窗模特设计

绘画主体
Window mannequin dressed in women's tennis uniform, ready to serve, holding racket in hand, wearing white sneakers
橱窗模特穿着女子网球服，准备发球，手里拿着球拍，穿着白色运动鞋

场景
Smooth cut, simple design, dynamic, full of sunlight, blurred background, front view, smooth cut, expensive luxury, super detailed details, perspective, through the glass
光滑剪裁，简单设计，动感，充满阳光，模糊背景，正视图，光滑剪裁，昂贵奢华，超详细细节，透视，透过玻璃

风格
Simple sporty style
简约的运动风

画质
Real photography, 32K quality
真实摄影，32K 画质

基础设置
--v 5.2
版本 V5.2

例 3. 服装材质设计

绘画主体

Women's V-neck suspender evening dress design, satin material, matte, V-neck, silver sequins, hip-covering skirt, pearl decoration
女式 V 领吊带晚礼服设计，缎面材质，哑光，V 领，银色亮片，包臀裙摆，珍珠装饰

场景	风格
Supermodel body proportions, bright light 超模身材比例，光线明亮	Retro style, clothing design 复古风格，服装设计

画质	基础设置
Real photography, 32K quality 真实摄影，32K 画质	--v 5.2 版本 V5.2

例 4. 服装风格设计

绘画主体

Rainbow jacket, liquid emulsion print style, light yellow and dark teal, digital print, dark orange and dark teal

彩虹夹克，液体乳液印花风格，浅黄和深青色，数码印花，深橙色和深青色

场景

Free flowing lines, geometric inspiration, advertising studio lighting, detailed clothing design

自由流动的线条，几何灵感，广告工作室灯光，详细的服装设计

风格

Yohji Yamamoto style
山本耀司风格

画质	基础设置
Real photography，32K quality 真实摄影，32K 画质	--ar 3:4 --v 5.2 画面比例 3:4 版本 V5.2

例 5. 汉服元素服装设计

绘画主体

Hanfu clothing, clothing design, Song Dynasty, Hanfu, light green and white, red-crowned crane and landscape patterns, cotton and linen material, embroidery
汉服服装，服装设计，宋代，汉服，浅绿白，丹顶鹤和山水图案，棉麻材质，刺绣

场景

Clothes display, supermodel proportions, intricate details, studio background, high key lighting, detailed details
服装展示，超模比例，精致复杂的细节，工作室背景，高调光线，详细的细节

风格

Ancient Chinese style
中国古代风格

画质

Real photography, super high definition
真实摄影，超级高清

基础设置

--v 5.2
版本 V5.2

例 6. 羽绒服服装设计

绘画主体
Various sketches and ideas about down jackets, white snow mountain camouflage
关于羽绒服的各种草图和想法，白色雪山迷彩

场景
Expert drawing, digital painting style, detailed style, precise details, concept art
专家绘图，数字绘画风格，风格详细，细节精准，概念艺术

风格	画质	基础设置
Clothing design, simplicity	Ultra-detailed rendering, 4K quality	--v 5.2
服装设计，简约	超细致渲染，4K 画质	版本 V 5.2

例7. 男士毛衣设计

绘画主体

Men's striped cardigan sweater, boys' sweater, loose fit, available in green and beige, dark sky blue and dark beige, light green and light black

男式条纹开衫毛衣，男童毛衣，宽松版型，有绿色和米色、深天蓝色和深米色、浅绿色和浅黑色可供选择

场景

Playful details, stripes and shapes, decorative details and pockets, front view, ambient light

俏皮细节，条纹和形状，装饰细节和口袋，正视图，环境光

风格
Advertising photography style
广告摄影风格

画质
Real photography, HD
真实摄影，高清

基础设置
--v 5.2
版本 V5.2

例 8. 女士连衣裙设计

绘画主体

Women's skirt designs, wear on mannequin, dress, square neck, ruffle skirt, midi skirt,
女式裙子设计，穿在人体模型上，连衣裙，方领，荷叶边裙子，中长裙

场景

Complex pattern design, fresh pastoral style, soft light
复杂图案设计，清新田园风格，柔和的光线

风格

Costume design, comic style
服装设计，漫画风格

画质	基础设置
32K quality	--niji 5
32K 画质	版本 Niji5

例 9. 男士西装设计

绘画主体

Men's suit, striped design, gray and black, lapel collar
男士西装，条纹设计，灰黑色，翻领

场景

Gentlemanly atmosphere, classic style, product view, frontal panorama, realistic photography, dark background, advertising lighting, meticulous details
绅士气息，经典款式，产品视图，正面全景，写实摄影，深色背景，广告灯光，细致细节

风格

Simple and elegant, elegant gentleman style
简约大方，优雅的绅士装风格

画质

super high quality, realistic photography, 32K quality
超高品质，写实摄影，32K 画质

基础设置

--v 5.2
版本 V5.2

例 10. 女士水桶包设计

绘画主体
Bucket bag, leather, gray, shoulder bag
水桶包，皮质，灰色，单肩包

场景
Simple design, high quality, casual style, commuting scene, bright light, fashionable and generous, austere and minimalist
简约设计，高品质，休闲风格，通勤场景，光线明亮，时尚大方，大气简约

风格
Simple clothing design
简约的服饰设计

画质
Realistic photography, street photography, 32K quality
写实摄影，街头摄影，32K 画质

基础设置
--v 5.2
版本 V5.2

例 11. 男士邮差包设计

绘画主体
Men's crossbody bag, leather and canvas combination, camouflage dark pattern, multifunctional pocket design
男士斜挎包，皮革与帆布组合，迷彩深色图案，多功能口袋设计

场景
Commuting use, simple design, smooth cut, solid color background, global illumination, product view, front view
通勤使用，简约设计，剪裁流畅，纯色背景，全局照明，产品视图，正视图

风格
Simple clothing design
简约的服饰设计

画质
Photorealistic photography, 32K quality
照片写实摄影，32K 画质

基础设置
--v 5.2
版本 V5.2

例 12. 女士托特包设计

绘画主体

Women's handbag design, handbag, tote bag, large capacity, black, leather
女式手提包设计，手提包，托特包，大容量，黑色，皮革

场景

Placed on the cabinet, clean background, simple design, autumn atmosphere, bright light, golden section
放在柜子上，干净的背景，简约的设计，秋季气息，光线明亮，黄金分割

风格

Yayoi Kusama style, Chanel style
草间弥生风格，香奈儿风格

画质

Realistic photography
写实摄影

基础设置

--v 5.2
版本 V5.2

例 13. 男士皮鞋设计

绘画主体

A pair of distressed brown-gray formal shoes, dyed and washed, with hard edges
一双做旧棕灰色正装鞋，染色与水洗，硬边线条

场景

Oxford style, retro aesthetics, antique theme, nostalgic atmosphere, product view
牛津风，复古美学，仿古主题，怀旧气息，产品视图

风格	画质	基础设置
Retro Oxford shoe style	Real photography, HD	--v 5.2
复古牛津鞋风格	真实摄影，高清	版本 V5.2

例 14. 女士高跟鞋设计

绘画主体

Women's high heels, black and white, square diamond buckle, thick heel, square toe
女式高跟鞋，黑白，方钻扣，粗跟，方头

场景

French style, minimalist design, smooth cuts, global illumination, product view, side view
法式风情，简约设计，光滑剪裁，全局照明，产品视图，侧视图

风格	画质	基础设置
Audrey Hepburn style	Realistic photography, 32K quality	--v 5.2
奥黛丽·赫本风格	写实摄影，32K 画质	版本 V5.2

例 15. 女士高跟鞋材质设计

绘画主体
Women's high-heeled shoes design, corduroy material, burgundy and black, square diamond buckle, thick block heel, square toe
女式高跟鞋设计，灯芯绒材质，酒红和黑色，方钻扣，粗跟，方头

场景
French style, simple design, smooth tailoring, global lighting, product view, side view
法式大气，简约设计，流畅剪裁，全局照明，产品视图，侧视图

风格
Audrey Hepburn style
奥黛丽·赫本风格

画质
Realistic photography, 32K quality
写实摄影，32K 画质

基础设置
--v 5.2
版本 V5.2

例 16. 运动鞋设计

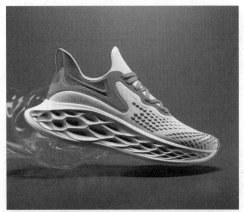

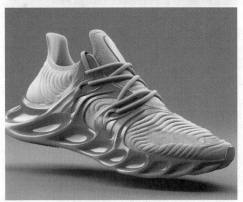

绘画主体

3D men's sports running shoes, lavender gradient, gorgeous geometric texture, spiral detailed pattern

三维男士运动跑鞋，淡紫色渐变，华丽的几何纹理，螺旋细致图案

场景

Ultra-fine details, clean backgrounds, global illumination, sporty atmosphere

超精细的细节，干净的背景，全局照明，运动氛围

风格

Master design style

大师设计风格

画质

Industrial design photography, 32K quality

工业设计摄影，32K 画质

基础设置

--v 5.2　版本 V5.2

例 17. 鞋子花纹设计

绘画主体

Complex pattern design, roses hidden among leaves, red, black and green
复杂的图案设计,隐藏在树叶间的玫瑰,红、黑和绿

场景

Graffiti texture, harmonious color matching, ultra-fine details, bright street atmosphere, fashionable atmosphere, side view
涂鸦质感,和谐的配色,超精细的细节,明亮的街头气息,时尚气息,侧视图

风格

Hip-hop
嘻哈风格

画质

Realistic photography, 32K quality
写实摄影,32K 画质

基础设置

--v 5.2
版本 V5.2

例 18. 童袜设计

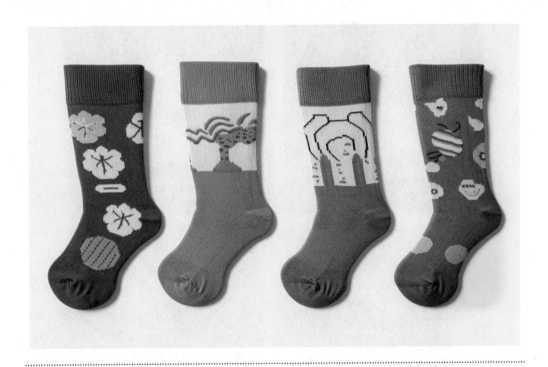

绘画主体

Four pairs of socks, flowers, cartoon characters, green, wood color, embroidery, wool material
四双袜子，花朵，卡通人物，绿色，原木色，刺绣，羊毛材质

场景

Strong sense of lines, simple design, earth color palette style, clean background, strong use of colors, multiple layers, simple set, wavy lines and organic shapes, side view
线条感强，设计简约，大地色板风格，干净的背景，色彩运用强烈，多层次，简约套装，波浪线和有机形状，侧视图

风格
Children's clothing design style
童装设计风格

画质
Product photography
产品摄影

基础设置
--ar 3:2 --v 5.2
画面比例 3:2 版本 V5.2

例 19. 童鞋（运动鞋）设计

绘画主体

A pair of children's sneakers, rubber sole, round toe, elastic buckle laces, breathable mesh cotton, lilac and pink combination
一双儿童运动鞋，橡胶底，圆头，松紧扣鞋带，透气网棉，淡紫色和粉色组合

场景

Simple design, smooth tailoring, global illumination, product view, side view
简约设计，流畅剪裁，全局照明，产品视图，侧视图

风格	画质	基础设置
Simple children's shoe design style 简约的童鞋设计风格	Photorealistic photography, 32K quality 照片写实摄影，32K 画质	--v 5.2 版本 V5.2

例 20. 宠物衣服设计

绘画主体

Dog clothes, tutu, pink and white, lace cotton

小狗衣服，芭蕾舞短裙，粉白色，蕾丝材质

场景

Bright light, central composition, model photo, full body photo

光线明亮，中心构图，模特照，全身照

风格

Cute style

可爱风格

画质

Realistic photography, 4K quality

写实摄影，4K 画质

基础设置

--v 5.2

版本 V5.2

例 21. 家纺设计

绘画主体
Tropical floral print bed set, in colorful washed styles, in pale yellow and aquamarine tones
热带花卉印花床套装，彩色水洗风格，浅黄色和海蓝宝石色调

场景
Fluffy duvet, reactive prints and dyes, flat forms, warm bedroom atmosphere, bedding display, global lighting
蓬松的羽绒被，活性印花和染料，平面形式，温暖的卧室氛围，床上用品展示，全局照明

风格	画质	基础设置
Warm home style	Realistic photography, 32K quality	--v 5.2
温馨的家居风格	写实摄影，32K 画质	版本 V5.2

例 22. 地毯花纹设计

绘画主体
Oriental rug with floral pattern on the floor, light red and dark navy style, dark green and maroon
地板上有花卉图案的东方地毯，浅红色和深蓝色风格，深绿色和栗色

场景
Carefully designed with perfect lines, light brown and dark black, jacquard art, light blue and maroon, intricate filigree details, tangled delicate vines, flat surface
精心设计的完美线条，浅棕色和深黑色，提花艺术，浅蓝色和栗色，复杂的花丝细节，缠绕精致藤蔓，平坦表面

风格
luxury, ultra-detailed, exquisite craftsmanship
豪华，超细节，精湛工艺

画质
Realistic photography, 32K quality
写实摄影，32K 画质

基础设置
--ar 3:4 --v 5.2
画面比例 3:4   版本 V5.2

例 23. 珠宝设计

绘画主体

The olive leaf-shaped crown is composed of green stones and diamonds
橄榄叶形皇冠由绿色宝石和钻石组成

场景

Jewelry design, royal jewelry atmosphere, exquisite craftsmanship, ultra-fine details, high quality, asymmetrical design, clean background, bright light, vintage atmosphere
珠宝设计，皇家珠宝氛围，精湛的工艺，超精细的细节，高品质，不对称设计，干净的背景，光线明亮，复古气息

风格	画质	基础设置
Modern jewelry style 现代珠宝风格	32K quality 32K 画质	--v 5.2 版本 V5.2

例 24. 模特珠宝展示设计

绘画主体

Jewelry display model, wearing emerald ring and silver bracelet, gently dragging her face, wearing black dress, showing jewelry pose
珠宝展示模特，戴着祖母绿戒指和银手镯，轻轻托着脸，身穿黑色连衣裙，展示珠宝的姿势

场景

Jewelry design display, real studio shooting, bright and elegant atmosphere
珠宝设计展示，摄影棚实拍，明亮和优雅的氛围

风格	画质	基础设置
Fashion advertising photography style 时尚广告摄影风格	HD 高清	--v 5.2 版本 V5.2

例 25. 运动手表设计

绘画主体

Smart sports bracelet design, LCD display, cyan and white
智能运动手环设计，液晶显示屏，青色和白色

场景

Metal texture, industrial design, light background, central composition, highlighting product texture, attracting visual appeal, advanced lighting
金属质感，工业设计，浅色背景，中心构图，凸显产品质感，吸引视觉注意力，高级照明

风格

Simple style, simple streamlined design
简约风格，简约流线型设计

画质

UHD rendering, 8K quality, realistic photography
超高清渲染，8K 画质，写实摄影

基础设置

--v 5.2
版本 V5.2

例 26. 女士手表设计

绘画主体

Women's watch, square dial model, white dial, brown embossed strap, glass mirror
女表，方形表盘款，白色表盘，棕色压花表带，玻璃镜面

场景

Simple design, exquisite simplicity, close-up, ultra-fine details
简约设计，精致简约，特写，超精细的细节

风格

Simple watch design
简约的手表设计

画质

Product photography, 8K quality
产品摄影，8K 画质

基础设置

--v 5.2
版本 V5.2

例 27. 户外充电器设计

绘画主体

Outdoor battery design, portable generator, high power, large capacity, 1000W, portable handle, solar charging panel, rich interfaces, black and gray

户外电池设计，便携式发电机，大功率，大容量，1000W，便携式手柄，太阳能充电板，接口丰富，黑色和灰色

场景

On stones by the stream, natural camping atmosphere, asymmetrical composition, front view, good light, good depth of field effect

在小溪边的石头上，自然露营氛围，不对称构图，正视图，光线好，景深效果好

风格

Industrial design style
工业设计风格

画质

Real photography, 32K quality
真实摄影，32K 画质

基础设置

--v 5.2
版本 V5.2

例 28. 户外登山包设计

绘画主体

Outdoor backpack mountaineering backpack design, orange mountaineering backpack, gray shoulder straps, gray zipper, light silver and orange, with compartments, with handles, solid color
户外背包登山背包设计，橙色登山背包，灰色肩带，灰色拉链，浅银和橙色，带隔层，带手柄，单色

场景

Capable, precise, dynamic outdoor shooting, suitable for the terrain of northern China
干练，精准，动态户外拍摄，适合中国北方地形

风格	画质	基础设置
Swiss outdoor style	Product photography, high-definition, 4K quality	--v 5.2
瑞士户外风格	产品摄影，高清，4K 画质	版本 V5.2

例 29. 户外帐篷设计

绘画主体

Light indigo and light amber tent on the lawn, mushroom shape, canvas material
浅靛蓝和浅琥珀色的草坪帐篷，蘑菇形状，帆布材质

场景

Organic materials, vibrant colors, soft styles, childlike atmosphere, dreamlike quality, sharp and vivid, pure colors, nostalgic atmosphere, morning light
有机材质，充满活力的配色，柔和的风格，童趣氛围，梦幻般的品质，锐利生动，色彩纯正，怀旧气氛，晨光

风格

International style, hiking
国际风格，徒步旅行

画质

Photorealistic photography, 32K quality
照片写实摄影，32K 画质

基础设置

--v 5.2
版本 V5.2

例 30. 餐具设计

绘画主体

Lotus leaf-shaped dinner plate with clear veins, dark and light green, irregular bowl mouth, curved and irregular style, geometric shapes and patterns and naturally inspired compositions, fresh colors, decorative details and embellishments
叶脉清晰的荷叶形餐盘，深浅绿色，不规则碗口，曲线不规则款式，几何形状和图案与自然灵感的构图，色彩清新，装饰细节和点缀

场景

Centered composition, global illumination, light background
居中构图，全局照明，浅色背景

风格

Simple design, elegant style, charming light Chinese style, ceramic texture
设计简洁，风格典雅，迷人的轻中国风，陶瓷质感

画质

4K quality
4K 画质

基础设置

--v 5.2
版本 V5.2

例 31. 酸奶包装盒设计

绘画主体
A yellow peach flavored yogurt packaging box
一款黄桃味酸奶包装盒

场景
Highlighting the natural theme, Monet's atmosphere, placed on the yellow peach tree with a spring atmosphere, bright light, asymmetrical composition
突出自然主题，莫奈氛围，置于春日气息的黄桃树上，光线明亮，不对称的构图

风格
Product packaging design
产品包装设计

画质	基础设置
Realistic photography 写实摄影	--v 5.2 版本 V5.2

例 32. 产品系列化设计

绘画主体

Business stationery display, pens, envelopes, notebooks and other stationery, light gray, solid color artwork

商业文具展示，钢笔、信封、笔记本等文具，浅灰色、纯色艺术品

场景

For commercial display purposes

商用展示用途

风格

Simple design

简约设计

画质

UHD images, 32K quality

超高清图像，32K 画质

基础设置

--v 5.2

版本 V5.2

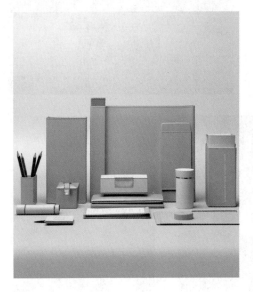

例 33. 儿童闹钟设计

绘画主体

Children's alarm clock, apple shape, mainly green, placed on the desk
儿童闹钟，苹果形状，以绿色为主，放在书桌上

场景

Ultra-high details, childlike and harmonious color design, learning atmosphere, background blur, morning sunlight, front view
超高细节，童趣和谐的色彩设计，学习氛围，背景虚化，早晨阳光，正视图

风格

Product display
产品展示

画质

Realistic photography, 8K quality
写实摄影，8K 画质

基础设置

--v 5.2
版本 V5.2

例 34. 儿童车设计

绘画主体

Blue children's car, light blue and light pink, plastic texture, detailed spiral pattern on the car, ocean theme
蓝色儿童车，浅蓝色和浅粉色，塑料质感，车上螺旋细致图案，海洋主题

场景

Childlike texture atmosphere, clean background
童趣质感氛围，干净的背景

风格	画质
Salon style	Realistic photography, 32K quality
沙龙风格	写实摄影，32K 画质

基础设置

--ar 8:6 --v 5.2
画面比例 8:6　版本 V5.2

例 35. 唱片机设计

绘画主体
Vinyl record player, neon colors
黑胶唱片机，霓虹色

场景
Transparent texture, global illumination, fashion home background, product view, frontal panorama, light background, advertising lighting
透明质感，全局照明，时尚家居背景，产品视图，正面全景，浅色背景，广告灯光

风格
Modern industrial design, simple style
现代工业设计，简约风格

画质
Fine details, ultra-high image quality, realistic photography, 32K quality
精细细节，超高画质，写实摄影，32K 画质

基础设置
--v 5.2
版本 V5.2

例 36. 耳机设计

绘画主体
Headphones, plant pattern, enamel material
耳机，植物图案，珐琅材质

场景
Bright light, luster, product design
光线明亮，光泽感，产品设计

风格
Retro style
复古风格

画质
Realistic photography, HD
写实摄影，高清

基础设置
--v 6.0
版本 V6.0

例 37. 猫窝设计

绘画主体
Pet house for orange and blue birds, dark gray and light beige style
橙色和蓝色小鸟的宠物屋，深灰色和浅米色风格

场景
Playful, cute and colorful, interesting patterns, bright colors, soft fabrics, warm atmosphere, diverse patterns, functional aesthetics
俏皮，可爱多彩，有趣的图案，明亮的色彩，柔软的面料，温馨的氛围，图案多样，功能美学

风格
Childlike style
充满童趣的风格

画质	基础设置
Real photography 真实摄影	--v 5.2 版本 V5.2

例 38. 换鞋凳设计

绘画主体
Shoe-changing stool design, cute sheep shape, plush material, wooden stool legs
换鞋凳设计，可爱的羊造型，毛绒材质，木质凳腿

场景
Elegant Morandi color scheme, childlike style, light background, bright light
优雅的莫兰迪配色，童趣风格，浅色背景，光线明亮

风格
Simple home design
简约家居设计

画质
Realistic photography, 8K quality
写实摄影，8K 画质

基础设置
--v 5.2
版本 V5.2

例 39. 化妆品套装设计

绘画主体

The skincare range consists of a variety of products and two boxes, with branding and packaging incorporating symbolic natural elements such as icebergs and spring water, the main colors are white, gray and green
该护肤产品系列包含多种产品和两个盒子,其品牌和包装结合了冰山和泉水等象征性自然元素,主要颜色为白灰色和绿色

场景

Harmonious, elegant and organic flow of color combinations, lines, vibrant energy, light background
和谐、优雅、有机流动的色彩组合,线条,充满活力的能量,浅色背景

风格

Featuring Art Nouveau style, photorealistic rendering, full-tone printing
具有新艺术风格,真实感渲染,全色调打印

画质

Product advertising photography, 8K quality
产品广告摄影,8K 画质

基础设置

--v 5.2
版本 V5.2

例 40. 咖啡机设计

绘画主体

Industrial design, coffee machine, combination of retro wood and metal, simple design

工业设计，咖啡机，复古木头和金属的结合，简约的设计

场景

High-end texture, retro texture, ultra-fine details, placed in a bright cafe, global illumination, depth of field effect

高端的质感，复古的质感，超精细的细节，放在明亮的咖啡馆里，全局照明，景深效果

风格

Retro style

复古风格

画质

Product photography, 32K quality

产品摄影，32K 画质

基础设置

--v 5.2

版本 V5.2

例 41. 饼干包装铁盒设计

绘画主体
A square tin box for biscuits, it has a bunny and a small house motif
一个装饼干的方形锡盒子,上面有一只兔子和一个小房子的图案

场景
Product packaging design, brand new, no handle
产品包装设计,全新,无把手

风格
Peter Rabbit style, Illustration
彼得兔风格,插图

基础设置
--ar 3:4 --v 6.0
画面比例 3:4　版本 V6.0

例 42. 蓝牙音箱设计

绘画主体
Bluetooth speaker, white and gray, drop-shaped shape
蓝牙音箱，白色和灰色，水滴形造型

场景
Simple design, cool style, the background is a simple wooden house, light background, depth of field effects
简约设计，冷酷风格，背景是简单的木屋，浅色背景，景深效果

风格
Minimalist home design
简约的家居设计

画质
Professionally shot through product photography
通过产品摄影专业拍摄

基础设置
--v 5.2
版本 V5.2

例 43. 破壁机设计

绘画主体
Wall breaking machine, transparent container, light yellow body
破壁机，透明容器，浅黄色机身

场景
Soft lines, bright kitchen atmosphere, placed on the kitchen island, depth of field effect, professional product photography
线条柔和，明亮的厨房氛围，放置在厨房岛台，景深效果，专业产品摄影

风格
furniture design
家居设计

画质
Realistic photography
写实摄影

基础设置
--v 5.2
版本 V5.2

例 44. 剃须刀设计

绘画主体

Razor design, black and dark gold, metallic texture
剃须刀设计，黑色和暗金色，金属质感

场景

Ergonomic design, high-end texture, ergonomic smooth curves, high-quality materials, light background, central composition, highlighting product texture, attracting visual appeal, advanced lighting
人体工学设计，高端质感，人体工学流畅曲线，优质材料，浅色背景，中心构图，凸显产品质感，吸引视觉吸引力，高级照明

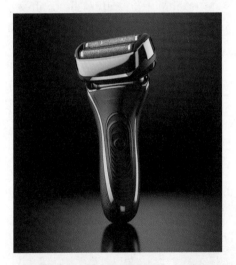

风格

Simple style
简约风格

画质

Product photography, UHD rendering, real photography
产品摄影，超高清渲染，真实摄影

基础设置

--v 5.2
版本 V5.2

例 45. 吸尘器设计

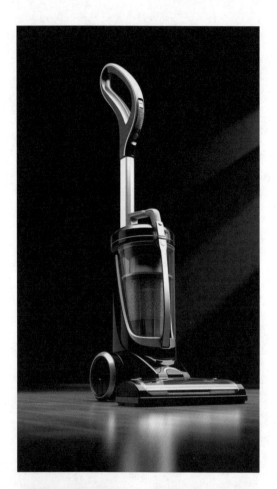

绘画主体

Vacuum cleaner design, black transparent glass, metal texture
吸尘器设计，黑色透明玻璃，金属质感

场景

Industrial design, simple streamlined design, clean living room, central composition, depth of field effect, highlighting product texture, attracting visual appeal, advanced lighting
工业设计，简单的流线型设计，干净的客厅，中心构图，景深效果，突出产品质感，吸引视觉吸引力，高级照明

风格

Simple industrial design style
简约的工业设计风格

画质

Realistic photography, UHD rendering, 8K quality
写实摄影，超高清渲染，8K画质

基础设置

--ar 3:5 --v 5.2
画面比例 3:5　版本 V5.2

例 46. 适老化轮椅设计

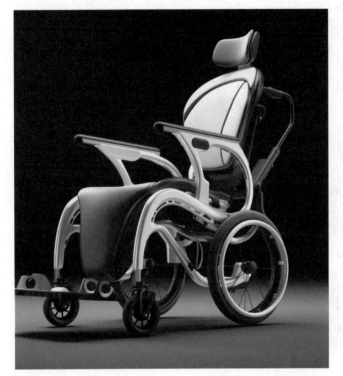
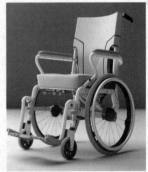

绘画主体

Wheelchair design, white and light silver
轮椅设计，白色和浅银色

场景

Modular design, use effect, elegance, volumetric lighting style, light background
模块化设计，使用效果，优雅感，体积灯光风格，浅色背景

风格	画质	基础设置
The chair adopts an age-appropriate design, suitable for the elderly, ergonomic design 椅子采用适龄设计，适合老年人，人体工程学设计	UHD images, advertising photography, 8K quality 超高清图像，广告摄影，8K 画质	--v 5.2 版本 V5.2

例 47. 展厅设计

绘画主体

The Expo booth with dimmed lights, exquisite woodblock print style, light white and light brown
灯光调暗的世博展台，木版画风格精致，浅白浅棕

场景

Pure colors, strong expressions, minimalist brush lines, minimalist black and white
色彩纯正，表情强烈，笔线极简，极简黑白

风格	画质	基础设置
Advertising style 广告风格	HD rendering, 8K quality 高清渲染，8K 画质	--v 5.2 版本 V5.2

第 11 章

室内设计

例1. 厨房设计

绘画主体
Kitchen design real effect, 8 square meters
厨房设计实景效果，8 平方米

场景
Reasonable layout, natural light, clever use of color, bold use of contrasts
布局合理，自然光，巧妙运用色彩，大胆运用对比

风格
Art deco flair style, neo-traditional, design by Kelly Wearstler, extreme details
装饰艺术风格，新传统，由凯莉·韦斯特勒（美国著名室内设计师）设计，极致的细节

画质
3d rendering, 32K quality
三维渲染，32K 画质

基础设置
--ar 5:3 --v 5.2
画面比例 5:3　版本 V5.2

例 2. 儿童书房设计

绘画主体
A study room with chair and table, about 8 square meters, designed for children
一间带桌椅的书房,约 8 平方米,专为儿童设计

场景
Panorama, comfortable atmosphere
全景,舒适的氛围

风格
Postmodern architecture and design, hybrid of contemporary and traditional, perfect details
后现代建筑与设计,现代与传统的融合,完美的细节

画质
8K quality, 3D
8K 画质,三维

基础设置
--ar 5:3 --v 5.2
画面比例 5:3   版本 V5.2

例 3. 衣帽间设计

绘画主体
Women's cloakroom, with mirror, about 6 square meters
女士衣帽间，带镜子，约 6 平方米

场景
Global illumination, well-structured
全局照明，结构良好

风格
Neoclassical style, exquisite craftsmanship, simple and elegant, home, daily
新古典主义风格，精湛工艺，简约大方，家居，日常

画质
16K resolution
16K 分辨率

基础设置
--ar 5:3 --v 5.2
画面比例 5:3　版本 V5.2

例 4. 卧室设计

绘画主体
A bedroom with a bed and a dresser
一间配有床和梳妆台的卧室

场景
Eye-level, serene and peaceful ambiance
水平视角，宁静祥和的氛围

风格
Modern minimalist style, pastel color, light emerald and beige
现代简约风格，柔和的色彩，浅翠绿和米色

画质
Rendered in Cinema 4D
在 Cinema 4D 中渲染

基础设置
--ar 5:3 --v 5.2
画面比例 5:3　版本 V5.2

例5. 大门设计

绘画主体
Ornate wrought iron arched door, dark silver and gold, luxurious texture
华丽的铁艺拱形门，暗银和金色，奢华的纹理

场景
Front view, Gilded Age, peaceful garden view, detailed rendering, luxurious textures, bright light, ultra-detailed details, clean background
正视图，镀金时代，宁静的花园景观，细致渲染，奢华的纹理，光线明亮，超细致的细节，干净的背景

风格
Rococo elegance, neoclassical symmetry, realistic style
洛可可优雅，新古典对称，写实风格

画质
Realistic, ultra-detailed rendering style, rendered in photorealistic style, 32K quality
真实摄影，逼真的超详细渲染风格，以照片写实风格渲染，32K 画质

基础设置
--ar 5:3 --v 5.2
比例 5:3 版本 V5.2

例 6. 儿童房设计

绘画主体

A cute children's bedroom, with a bed and study desk, carpet, throw pillows, wardrobe
一间可爱的儿童卧室，配有床和书桌、地毯、抱枕、衣柜

场景

Medium shot, incandescent lighting
中景，白炽灯

风格

Fairy tale theme, whimsical subject matter, sense of reality
童话题材，异想天开的题材，真实感

画质

Rendered in Maya
在 Maya 中渲染

基础设置

--ar 5:3 --v 5.2
画面比例 5:3　版本 V5.2

例 7. 茶室设计

绘画主体

A tea room, a Chinese tea table, white gauzy curtains, hemp rope-woven cushions
茶室，中式茶几，白色纱帘，麻绳编织坐垫

场景

Medium close-up, bright tone
中等特写，明亮的色调

风格

Simple style, hyper-realistic
简约风，超现实主义

画质	基础设置
8K quality	--ar 3:2 --v 5.2
8K 画质	画面比例 3:2　版本 V5.2

例 8. 中式餐厅设计

绘画主体
A Chinese inspired dining room
一个中式风格的餐厅

风格
In the style of soft atmospheric scenes, mori kei, realistic style
柔和大气的风格，森系，写实风

画质
8K quality
8K 画质

基础设置
--ar 2:3 --v 5.2
画面比例 2:3　版本 V5.2

例 9. 绿植角设计

绘画主体
Plant corner, green plant
植物角落，绿植

场景
Balcony corner
阳台角落

画质 | 基础设置
8K quality | --ar 2:3 --v 5.2
8K 画质 | 画面比例 2:3 版本 V5.2

例 10. 展示大厅设计

绘画主体
An ancient and mysterious palace hall
古老而神秘的宫殿

场景
Golden light, gigantic scale, front view, ultra wide angle, ray tracing
金色的光，巨大的规模，正视图，超广角，光线追踪

风格
In the style of Rococo-inspired art, realistic and hyper-detailed renderings
洛可可风格，逼真和超细致的渲染

画质	基础设置
16K quality 16K 画质	--v 5.2 版本 V5.2

例 11. 家庭影院设计

绘画主体

A home theater, with a large screen and some chairs, beautiful and modern
一个家庭影院,有一个大屏幕和一些椅子,美丽而现代

场景

Intense and dramatic lighting, precise simplicity, natural lighting, elevation perspective, panorama
强烈而戏剧性的灯光,精确简洁,自然采光,仰角透视,全景

风格

American style, use of precious materials, meticulous design, very atmospheric, surreal
美式风格,采用名贵材料,精心设计,非常大气,超现实

画质
UHD quality
超高清画质

基础设置
--ar 5:3 --v 5.2
画面比例 5:3　版本 V5.2

例 12. 卫生间设计

绘画主体

A bathroom, with potted plants, highly detailed leaves, green, about 8 square meters
一间浴室，盆栽植物，叶子细节丰富，绿色，约 8 平方米

场景
First-person view
第一人称视角

风格
Light luxury style, neat and tidy, realistic, detailed rendering
轻奢风格，整洁，逼真，细致渲染

画质
8K photography
8K 摄影

基础设置
--ar 5:3 --v 5.2
画面比例 5:3　版本 V5.2

例 13. 铁艺床设计

绘画主体
Wrought iron bed
铁艺床

场景
Bedroom, beige tone
卧室，米白色色调

风格
Simple style
简约风

画质
8K quality
8K 画质

基础设置
--ar 2:3 --v 5.2
画面比例 2:3 版本 V5.2

例 14. 鞋柜设计

绘画主体

Curved shoe cabinet design, curved design, wooden shoe cabinet, rattan cabinet door, walnut, curved design

弧形鞋柜设计，弧形设计，木质鞋柜，藤编门，胡桃木，弧形设计

场景

Bright light

光线明亮

风格

Simple style, Gaudí design style, smooth curves

简约风格，高迪设计风格，流畅的曲线

画质

Realistic photography, HD

写实摄影，高清

基础设置

--v 6.0

版本 V6.0

例 15. 沙发设计

绘画主体

3D view of living rooms in contemporary, sofa
当代客厅的三维视图，沙发

场景

High angle view, minimalist backgrounds, realistic usage of light and color
高视角，极简背景，真实的光线和色彩运用

风格

Classic Japanese simplicity, best masterpiece
经典日式简约，最佳杰作

画质

3D rendering, high resolution
三维渲染，高分辨率

基础设置

--ar 5:3 --v 5.2
画面比例 5:3　版本 V5.2

例 16. 书桌设计

绘画主体
Desk design, desk, office desk, furniture
书桌设计，桌子，办公桌，家具

风格
Simple style, IKEA style
简约风格，宜家风格

场景
Beige color
米色

画质
8K quality
8K 画质

基础设置
--ar 2:3 --v 5.2
画面比例 2:3　版本 V5.2

例 17. 书柜设计

绘画主体		场景
Chinese wooden bookcase, round shape, exquisite details, retro elements, bookcase design 中式木质书柜，圆形造型，精致的细节，复古元素，书柜设计		Bright light 光线明亮

风格	画质	基础设置
Chinese style 中国风	Realistic photography 写实摄影	--v 6.0 版本 V6.0

例 18. 台灯设计

绘画主体
Table lamp
台灯

场景
Colorful, design by Cecilie Manz
彩色，由 Cecilie Manz 设计

风格
Ultra realistic style
超逼真写实风

画质
8K quality
8K 画质

基础设置
--v 5.2
版本 V5.2

例 19. 衣柜设计

绘画主体

Wooden wardrobe, intricate rose pattern, exquisite details, retro elements, wardrobe design

木质衣柜，复杂的玫瑰图案，精致的细节，复古元素，衣柜设计

场景	风格
Bright light	French retro style
光线明亮	法式复古风格

画质	基础设置
Realistic photography	--v 5.2
写实摄影	版本 V5.2

例 20. 瓷砖设计

绘画主体
Tile design, flooring, tile, blue and gold details, ceramic
瓷砖设计，地板，瓷砖，蓝色和金色的细节，陶瓷

场景
High-angle photography
高角度摄影

画质
8K quality
8K 画质

基础设置
--v 5.2
版本 V5.2

例 21. 阳台设计

绘画主体

A balcony, seats and small table, with flowers and plants
一个阳台,座位和小桌子,有花草

场景

External perspective, natural light, serene ambiance, soft sunlight
外部视角,自然光,宁静的氛围,柔和的阳光

风格

Mediterranean style, harmonious and comfortable, hyper-realistic
地中海风格,和谐舒适,超写实

画质	基础设置
8K photography	--ar 16:9 --v 5.2
8K 摄影	画面比例 16:9 版本 V5.2

例 22. 服装店铺设计

绘画主体
Cloth store, interior design, elegant clothing
服装店，室内设计，优雅的服装

场景
Light white and beige
浅白色和米色

风格
Simple style
简约风格

画质
8K quality
8K 画质

基础设置
--ar 3:2 --v 5.2
画面比例 3:2　版本 V5.2

例 23. 咖啡店铺设计

绘画主体
Coffee shop design, rock main body
咖啡店铺设计，岩石主体

场景
Natural light, warm atmosphere, rough
自然光，温馨氛围，粗犷的元素

风格
Interior design, original style
室内设计，原始风格

画质
Realistic photography
写实摄影

基础设置
--v 6.0
版本 V6.0

例 24. 主题快闪店设计

绘画主体

Pop-up shop, 3D architectural booth design, 30 square meters

快闪店,三维建筑展位设计,30 平方米

场景

Pink theme

粉色主题

风格

Fashion style

时尚风

画质

C4D OC rendering, 8K quality

C4D OC 渲染,8K 画质

基础设置

--ar 2:3 --v 5.2

画面比例 2:3　版本 V5.2

例 25. 移动咖啡车设计

绘画主体
Mobile coffee cart design, with coffee machine, green, usage scene is street
移动咖啡车设计，带咖啡机，绿色，使用场景为街道

场景
Bright light, side view, warm atmosphere
光线明亮，侧视图，温馨氛围

风格
Realistic photography, decoration design, highly detailed
写实摄影，装修设计，高细节

画质
16K quality
16K 画质

基础设置
--ar 4:3 --v 5.2
画面比例 4:3　版本 V5.2

例 26. 房车内部设计

绘画主体

Caravan design, caravan interior, caravan
房车设计，房车内部，房车

场景

Warm color
暖色调

画质

8K quality
8K 画质

基础设置

--ar 2:3 --v 5.2
画面比例 2:3　版本 V5.2

例 27. 书店设计

绘画主体
Bookstore design, train theme
书店设计，火车主题

场景
Natural light, warm atmosphere, train carriage elements
自然光，温馨氛围，火车车厢元素

风格
Interior design, modern style
室内设计，现代风格

画质
Realistic photography
写实摄影

基础设置
--v 6.0
版本 V6.0

例 28. 珠宝店设计

绘画主体
Jewelry store design, interior design, booth
珠宝店设计，室内设计，展台

风格
Fashion style
时尚风格

画质
8K quality
8K 画质

基础设置
--ar 2:3 --v 5.2
画面比例 2:3　版本 V5.2

例 29. 博物馆设计

绘画主体
Museum design, museum exhibition hall, glass display cases
博物馆设计，博物馆展厅，玻璃展柜

风格
Realistic style
写实风

画质
8K quality
8K 画质

基础设置
--ar 3:2 --v 5.2
画面比例 3:2 版本 V5.2

例 30. 新中式茶室设计

绘画主体

A modern Chinese tea room with traditional Chinese elements, oversized windows with sliding doors, chandeliers above the wooden tables emitting soft light, tatami mats on the floor, and potted plants outside the large glass door

一间带有中国风传统元素的现代中式茶室，带推拉门的超大窗户，木桌上方的吊灯发出柔和的灯光，地板上铺着榻榻米，大玻璃门外摆放着盆栽

风格	画质
Chinese style	32K quality
中国风	32K 画质

基础设置

--ar 3:4 -- v 6.0

出图比例 3:4 版本 V6.0

例 31. 水下餐厅设计

绘画主体
The under the sea restaurant
水下餐厅

场景
Dark cyan tone, tropical landscapes, lively coastal landscapes
深蓝色色调，热带景观，生动的沿海景观

风格
Minimalism
极简主义

画质
8K quality, high resolution
8K 画质，高分辨率

基础设置
--ar 2:3 --v 5.2
画面比例 2:3　版本 V5.2